Memorizing Pharmacology Mnemonics

Pharmacology Flashcards and Fill-Ins for the Future Nurse, Doctor, Physician Assistant, and Pharmacist

Tony Guerra, Pharm.D.
2018

Copyright © 2018 by Tony Guerra, Pharm.D.

ISBN: 978-1-387-82770-1

All rights reserved. No part of this book may be reproduced or transmitted in any form or by any means without written permission of the author.

Memorizing Pharmacology Mnemonics: Pharmacology Flashcards and Fill-Ins for the Future Nurse, Doctor, Physician Assistant and Pharmacist.

*To Mindy,
Brielle, Rianne, and Teagan*

TABLE OF CONTENTS

Author's note 1
  Overcoming anxiety and overwhelm 1
INTRODUCTION 2
  Why read this pharmacology book? 2
  Pain or puzzle? 3
  Mnemonics as memory glue 5
  How much pharmacology will stay in your brain? 6
  What's new about this book? 8
  How do we begin? 10
CHAPTER 1: GASTROINTESTINAL MNEMONIC FLASHCARDS 11
  1. Antacids – ACIDIC MEALS 13
  2. Histamine-2 receptor antagonists (H2RAs) – CALIFORNIA H_2 18
  3. Gynecomastia side effect drugs – FEMALE PARTS CARE 22
  4. Proton pump inhibitors (PPIs) – ULCER DEVELOPER 24
  5. Proton pump inhibitors (PPIs) – ULCER DEVELOPER SIDE EFFECTS 30
  6. Triple peptic ulcer disease (PUD) therapy – ACED PUD . 32
  7. Quad peptic ulcer disease therapies – BED-MIDNIGHT or MACE 35
  8. Disulfiram reaction – MEGA PUCED (PUKED) 38
  9. Prostaglandin, two prokinetics, and a protectant – MIMES STUCK IN STOMACH 41

10. Have Diarrhea? – BUILD A BARRIER..................46
11. Anticholinergic side effects – ABDUCT WATER..............49
12. Cholinergic side effects – SLUDGE WATER 51
13. Have Constipation? – PRE-DISPOSED.......................53
14. Have Nausea? – STOP ADD AFTER.........................57
15. Vomiting? – PROVIDE MAGAZINE........................63
16. Irritable bowel disease and irritable bowel syndrome – GI MISLED 66

CHAPTER 2: MUSCULOSKELETAL MNEMONIC FLASHCARDS 71
17. Non-Steroidal Anti-Inflammatory Drugs - NSAIDS 73
18. NSAIDs – PAINN MEDS KICK IN 75
19. Acetaminophen analgesics – ABATE HEADACHE..........81
20. CII Opioids – MY MY MY FREAKING HEAD HURTS OUCH OUCH 84
21. Morphine side-effects - MORPHINE 89
22. CIII / CIV Opioids – A/C TRAM WRECK....................91
23. Opioid Antagonists – OMNIBUS AGAINST OPIOIDS ...94
24. Opioid antagonists for Opioid-Induced Constipation (OIC) – THREE MEDICINALS 97
25. 5-HT$_1$ Agonists – FEARS ZIN HEADACHES..................99
26. Disease Modifying Anti-Rheumatic Drugs (DMARDS) – Non-Biologics – MASH FOR JOINTS 102
27. DMARDs - Biologics – GO ADD IN A BETTER DMARD 105
28. Bisphosphonates – FRAZIL BONES 108
29. Muscle Relaxants – CAR MET CYCLE = BACK TO MD 112
30. Anti-Gout Agents – PACIFY PEG's TOE 116

Table of Contents

CHAPTER 3: RESPIRATORY MNEMONIC FLASHCARDS. 119
31. 1st Generation Antihistamines – ALLERGY ABCD RHYME 121
32. 2nd Generation Antihistamines – CALLED FOR NON-DROWSY 125
33. Decongestants – STUFFED UP PEOPLE 129
34. Intranasal Corticosteroids – BFF MOM KNOWS NOSE 133
35. Cough – BIG DARN COUGH 136
36. Oral Corticosteroids – NOT PUMPED STEROIDS 139
37. Steroid side effects, -LONE 142
38. Inhaled Corticosteroids – FUMBLE TO BREATHE 144
39. Beta-2 Agonists – ALL SAVE LUNGS 147
40. Alpha and beta agonism and antagonism – ALPHA BETA BOX 150
41. Beta-2 Agonists + Inhaled Corticosteroids – 7 Fs PROPEL BREATH MOVES 152
42. Muscarinic Antagonists – ANTIMUSCARINIC 155
43. Leukotriene inhibitors, Anti-IgE, Epi – Z'MORE RESPIRATORY DRUGS 159

CHAPTER 4: IMMUNE MNEMONIC FLASHCARDS 161
44. Penicillins – TWO PENICILLINS AND TWO AMINOPENICILLINS 163
45. Penicillinase Resistant Penicillins –LACTAM (ACED) NO MORE 167
46. Cephalosporin Generations – 1) CEPH ONE 2) OR OX 3) TRI DIN TAZ 4) PIME 5) TAROL 171
47. Monobactam – AZTREONAM ALONE 176
48. Carbapenems – MEDIC 178
49. Glycopeptide – VANCOMYCIN RED HOMES 181

50. Vancomycin related lipoglycopeptides – TRIPOD 185
51. Time vs. Concentration Dependent – FRAMED CONCENTRATION CHART 188
52. Fluoroquinolones – MEDICAL 192
53. Aminoglycosides – GRAM NEGATIVES 196
54. Aminoglycosides MOA and side effects – AMINO 200
55. Bacteriostatic – STATIC 202
56. Sulfamethoxazole / Trimethoprim – MOA and Side Effects – SULFA 205
57. Oxazolidinones – MELT MRSA 208
58. Macrolides – FACE ATYPICAL BACTERIA 210
59. Tetracyclines – TIMED STAINED TEETH 214
60. Antituberculosis Agents – RIPER 217
61. Hepatotoxicity – LIVER PAIN 220
62. Antifungals – ANTI-FUNGAL EVICTIONS 222
63. Antivirals – Herpes Simplex Virus & Varicella Zoster Virus (HSV/VZV) – SAVED PROM 227
64. Antivirals – Influenza – FLU ZOMBIE 230
65. HIV antivirals – THEM THREE, RAID HIV 233
66. Antiviral – Respiratory Syncytial Virus – PEDIATRIC PALIVIZUMAB FOR PREVENTION 237
67. Three HEP-C Virus (HCV) Meds – DAILY DOSES 239

CHAPTER 5: NEURO/PSYCH MNEMONIC FLASHCARDS 241
68. Sedative/Hypnotics – A PEZZ PILLOW RESTTT 243
69. Anxiolytics - BE CALM with BUSPIRONE and BENZOS 249
70. Benzodiazepines – short to long-acting 253

Table of Contents

71. Extrahepatic benzodiazepine metabolism- OUTSIDE LOT 255
72. Eight antidotes – FIND AND PACK ANTIDOTES 257
73. Barbiturates – BARB SLEPT TOO LONG 259
74. Smoking Cessation – STOP SMOKING VIBE 262
75. ADHD Agents – CALMED ADHD 265
76. Depression diagnosis criteria – ESCAPISMS 270
77. Selective Serotonin Reuptake Inhibitors (SSRIs) – SSRI UPS AFFECT 272
78. Serotonin side effects (SSRIs) and serotonin syndrome symptoms (HARM) 277
79. Drugs causing depression – MOPS ALL DAY 280
80. Serotonin Norepinephrine Reuptake Inhibitors (SNRIs) – TWO HAPPY DVDS 282
81. Tricyclic and Tetracyclic Antidepressants (TCAs, TeCAs) – WITHOUT DOPAMINE 285
82. Tricyclic antidepressants – TCAS 289
83. Monoamine Oxidase Inhibitors (MAOIs) – PINOTS 291
84. Monoamine oxidase inhibitors for depression – THE PITS 294
85. Mood Stabilizers OCD – VOCALL MOOD 296
86. Carbamazepine – CARBA 299
87. Lithium: side effects – LITHIUM 301
88. 1st Generation Antipsychotics – NO POTENT CHATS. 304
89. Extrapyramidal symptoms (EPS) – NEVER ADAPT 308
90. 2nd Generation Antipsychotics – ABC OPQR & Z 310
91. Traditional Anti-Epileptics – PACED 314
92. Phenytoin: adverse effects PHENYTOIN 317
93. Newer Antiepileptics – LAPTOP LIGHTZ 319

v

Memorizing Pharmacology Mnemonics

94. Parkinson's Agents – SUBPAR CLUES TO CAUSE 323
95. Alzheimer's Agents – DREAMY GAL 327
96. Local anesthetics – BLOCK AXONS 330
CHAPTER 6: CARDIO MNEMONIC FLASHCARDS 333
97. Hypertension treatments – ABCD............................335
98. Renin angiotensin aldosterone system (RAAS) – AR AA AR AR..337
99. ACE Inhibitors – ACE BLOCKER QUICK FACTS..........340
100. ACEI side effects - CAPTOPRIL 344
101. Angiotensin II Receptor Blockers (ARBs) – VOCAL AID 346
102. Alpha$_1$ Blockers for Hypertension – BLOOD PRESSURE DIPPTS (DIPS) 349
103. Beta Blockers BY GENERATION – PLAN FOR EXAMS, GET COLA NOW 352
104. Beta Blockers' Selectivity – DO NOT USE: A THROUGH N, O THROUGH Z 357
105. Beta-blocker concerns - ABCDE............................359
106. Calcium Channel Blockers (CCBs) – Non-Dihydropyridines – SLOW DVD...361
107. Calcium Channel Blockers (CCBs) –Dihydropyridines – SAVED INFANTS..364
108. Diuretics – MAN, FLUSH THIS..............................367
109. Diuretic classes – LOCATE A BATHROOM SOON 371
110. Hydrochlorothiazide indications – HCTZ 373
111. Drug classes that increase potassium – POTASSIUM PANDA 375
112. Cholesterol lowering – HMG Co-A reductase inhibitors - STATINS – LDL FALLS SHARPLY................................377

vi

113. HMG-CoA reductase inhibitor (STATIN) details – HMG Co-A 381
114. PCSK9 Inhibitors – THE PCSK9 PEACH 383
115. Cholesterol-Lowering Agents – FIGHT ONCE AGAIN 385
116. Anticoagulants – HAS FEWER CLOTS 389
117. Warfarin: interactions –ACADEMIC'S FAB-4 394
118. Direct Oral Anticoagulants (DOACs) – EXPANDS BREADTH OF ANTICOAGULATION OPTIONS 396
119. Vasodilators – HINDER ANGINA 400
120. Antiplatelet Agents – ADD FOR PACT (packed) PLATELETS 403
121. STEMI drugs – MOAN 406
122. Emergency Drugs – LEAN AND DEMAND 408
CHAPTER 7: ENDOCRINE MNEMONIC FLASHCARDS.... 411
123. Insulins – LEARN INSULIN DRUGS 413
124. Biguanides and thiazolidinediones – ME DROP GLUCOSE 417
125. Sulfonylureas and a meglitinide – GLI GLI GLY GLUCOSE, RELEASE 420
126. GLP-1s and DPP-4s – LESS, LESS GLUCOSE 423
127. SGLT2 Inhibitors – EXCRETE GLUCOSE 426
128. Thyroid Medications – THYROID LAMP 428
129. Levothyroxine Narrow Therapeutic Index – TIE the TD / ED 431
130. Hormones and Birth Control – ESTR / GEST / STER (ESTER JUST STARES) 433
131. Urinary Incontinence Agents – LESS MODESTY NEEDED 437

132. Agents for Erectile Dysfunction –SAVE TONIGHT 440
133. Agents for BPH – FLOWED AT LAST............................444
134. Menopause Agents – COMPETE AGAINST
MENOPAUSE 447
Alphabetical list of stems..449
Generic and Brand Index..455

AUTHOR'S NOTE

OVERCOMING ANXIETY AND OVERWHELM

Imagine reading a pharmacology book that instead of causing anxiety and overwhelm, it reduces it. How could that happen?

Imagine someone wants a few things from the grocery store. If they speak slowly, only ask for three or four items, tell you exactly what aisle to look in, you're confident.

However, if someone speaks quickly, asks you to remember 13 or 14 items, doesn't know what aisle to look in, you feel anxious and overwhelmed. What can you do?

You can pick up a 3x5 card and make ordered categories from the left side of the store to the right. You say, "Okay, what fruits did you want?" "What frozen food did you need?" "What dairy or cheese did you ask for?" You classify and sort. You make a flash-card and you fill it in.

This book is the same. I've ordered pharmacology categories, from the left of the pharmacology store to the right. I've made flashcards, and created fill-ins. This book's system takes complex anxiety-causing and overwhelming amounts of information and sorts them quickly and memorably.

A future pharmacology student who previewed my book before she took my class said, "I feel better about it now." That's what I want for you.

INTRODUCTION

WHY READ THIS PHARMACOLOGY BOOK?

The quick answer is a question. Why learn pharmacology one drug at a time when I can teach you how to learn 4, 8, or 12 drugs at a time and remember them longer? But, let's start with a story about the real obstacles, a lack of energy and time.

Seven years ago, my wife and I were side-by-side in a hospital bed. We'd just heard she might deliver our triplet daughters at 19 weeks. The doctor gave her magnesium to stop the contractions, and she started vomiting from the effects. A nurse and nurse aid both walked in to help. It was around midnight and dark, but I saw both were prior pharmacology students of mine. They worked all night to comfort my wife through the magnesium toxicity.

In the morning, we got the news the doctor could perform a cerclage procedure to help keep the girls in until a much later gestation week. This mom, nurse, and student-nurse saved the day. But my wife, for her efforts, got 55 days of hospital bed rest. Bed rest is a misnomer; there is little rest.

She couldn't get comfortable to watch television or read. She could talk and listen. The best I could do was visit three times a day, before, during and after work. Visiting was also the most valuable act a friend could do for her, get her through just one more day as each day closer to 28 weeks gestation exponentially improved our daughters' survival chances.

INTRODUCTION

The student-nurse worked through the night, saving three lives and during daylight hours, she's exhausted. An instructor might think she doesn't care about the course as she yawns through it. Friends might think she doesn't care about them because she's sleeping or her erratic schedule keeps her missing important events. Classmates might wonder why she can't keep up with the projects and coursework. While I dedicate my books to my wife and daughters, I write these books for the future health professional, working through the night as a practitioner, parent, or both. I intentionally write them as a script, so if a student reads them as a book, it sounds like someone talking to them. If they listen to the audiobook, they can keep the information in mind even if they're dog tired.

PAIN OR PUZZLE?

Do you see pharmacology as a pain or a puzzle? I've found one question separates the successful and unsuccessful student. Students who see pharmacology as a pain, a threat to the grade point average and graduation, have no trouble finding people to agree. But they struggle through with little satisfaction.

Others see pharmacology as a puzzle, a challenging game to play alone or with a group. This book shows you many solutions to memorizing and applying puzzle pieces that don't fit at first. Writing or hearing a great mnemonic is like finding a corner piece that anchors the puzzle. In this book, I make pharmacology faster, easier, and more fun to learn, but also lay the solid groundwork to pull more knowledge from your test prep books and software.

1) **Faster.** When I read social media posts about the board exams, I find that those students that crushed the exam complete not one, but at least two test prep books. How? People that have more information memorized can move through test prep books faster. This book will help you build a stronger foundation.
2) **Easier.** Often, pharmacology professors suffer from the Curse of Knowledge. The material is too familiar. They fly through terms without realizing students aren't taking it all in. What a student needs the professor to do is talk "fewer." That is, a person can keep 3 to 7 items in short-term memory at any given time. I intentionally limited the mnemonics in this book to smaller numbers so you can better process the material.
3) **Fun.** While the future health professional's life is often deadly serious, the more you integrate an emotion, humor especially, into a mnemonic, the more likely you'll remember it.

A great mnemonic gives you a sense of purpose and helps you learn the material. For example, I use the mnemonic SAVED INFANTS to remember the dihydropyridine calcium channel blockers. At first, that may not make sense; calcium channel blockers are antihypertensives. But one use of nifedipine, the 'N' in INFANTS is to prevent uterine contractions. Nifedipine saved our triplet daughters' lives by helping them make it from a near delivery at week 19 to delivery at 27 weeks and three days.

Another mnemonic, SAVED TONIGHT, helps you remember what Viagra does, help couples re-engage. But also in the SAVED TONIGHT mnemonic, the 'N' for nitrous medications

INTRODUCTION

warns of a deadly interaction. Mnemonics work well when emotion and purpose combine.

MNEMONICS AS MEMORY GLUE

Have you read a pharmacology chapter only to forget it a minute later? If you're busy with work, other classes, or, like me, with 6-year-old triplet daughters, you don't have time for constant rereading. To keep from forgetting, you need memory glue – vibrant mnemonics that stick quickly and easily.

This book shrinks the work to remember the most essential information, including medication names, classifications, side effects, and mechanisms of action, into just over 130 slides. How do you know if you need this book? We can do a quick and dirty back-of-the-envelope test to see if this volume is for you.

If you're listening on your commute, name as many drug names as you can out loud. Number them as 'one, aspirin; two, ibuprofen; and so on' until you reach a number where you can't name more. If you're at a desk, write out a list of numbered drug names. Can you pass your final or board exam with that number? Would you feel better getting to 400 drug names?

What you'll find is that, at first, you quickly write out many drugs. Nevertheless, like popcorn that stops popping when only a few kernels remain, your ability to write drug names slows.

That's a dangerous place to be in if you need to answer questions immediately, as you will with board exams like the

NCLEX®, NAPLEX®, or USMLE ®. Most board exams won't let you revisit previous questions if you remember something after you've already passed a problem by, and what happens if you fail the boards? Well, let's make sure we don't find out.

Instead, look carefully at your drug name list. Did you group drug names? For example, if you wrote 'aspirin,' did you write other analgesics like ibuprofen and acetaminophen beside it? If you remembered penicillin, did you write another antibiotic like amoxicillin or think of the related infection, methicillin-resistant *Staphylococcus aureus* (MRSA)? Each group you have (or don't have) represents a short story your brain built. In this book, we intentionally create solid, memorable stories into a single flashcard slide.

There are three flashcard slide types; either a slide 1) contains only medication names, 2) contains only side effects, interactions, and mechanisms of action, or 3) it includes both. To provide contrast, I've bolded medication names if a slide contains both. Since you can't see them in an audiobook when reading slides, I read the medication's generic name first, then the brand name second.

How much pharmacology will stay in your brain?

Your mind has an infinite long-term memory capacity for organized information. However, sorting drugs into groups in pharmacology is like going to a different country. If you are using coins in the European Union, the coins make complete sense. The one-cent, five-cent, and ten-cent coins get

INTRODUCTION

progressively larger, are all bronze colored, and have the number one, five, or ten stamped on them.

As you progress from the 10-cent to 20-cent to 50-cent coins, the same logic follows. They are all gold colored, get progressively larger, and have a large 10, 20, or 50 stamped on them. That logical system makes remembering the coins' values easy.

However, the United States' currency follows a different reasoning. The first four coins in value order are a penny, nickel, dime, and quarter. Without reading English, you can't see that a penny is one cent, a nickel's value is five cents, and a quarter represents a quarter dollar. A dime reads "dime, d-i-m-e," giving no hint of its value. The coin's colors run copper, nickel, nickel, nickel. The size order from smallest to largest is the dime, penny, nickel, quarter or 10, 1, 5, and 25 cents, rather than 1, 5, 10, and 25 cents.

However, tourists who visit the United States for a few days learn the penny, nickel, dime, and quarter as the climbing value. To become familiar with pharmacology, in the same way, we need to spend time with the currency of pharmacology groups.

We can compare each drug class to a different country's currency. It's hard to sort foreign coinage. However, if we focus on only four items, like the penny, nickel, dime, and quarter, we can learn values quickly.

The value of working in this way is that you don't overload your working memory. You have a short story in your mind where you can quickly add new knowledge. Did you just read something about a proton pump inhibitor's side effect? You'll know that goes in the proton pump inhibitor story. If you've learned about a new indication for a beta-lactamase resistant penicillin antibiotic, you add that to the penicillins page.

WHAT'S NEW ABOUT THIS BOOK?

In *Memorizing Pharmacology: A Relaxed Approach*, I went through drug names one at a time. I reviewed the classification prefixes, suffixes, and mnemonics for each drug name. For example, I included the two H$_2$-blockers with the identical suffix –tidine in famo<u>tidine</u> and rani<u>tidine</u>. In this book, we step up the pace with multiple drugs within a class and unify them with a mnemonic.

When we focus on three to seven items, drugs, and side effects, the information remains within short-term cognitive memory limits. We can work with information faster and within the context of groups we can compare and contrast. With Bloom's taxonomy, a way to classify educational learning objectives, memorization is foundational.

With multiple drugs in memory, we can compare and contrast which is critical to understanding and reaching higher taxonomy levels. For example, when comparing cime<u>tidine</u> to rani<u>tidine</u>, we learn that ranitidine is better because of cimetidine's inhibitory effect on metabolic cytochrome P450 enzymes. We begin with a question on one page like this:

Question 2. Name four histamine-2 receptor blockers, their class suffix, two specific concerns about cimetidine, one about cimetidine and famotidine, and a general concern for the drug class itself.

While the question seems complicated, we can order our answer in a mnemonic flashcard, like this.

Histamine-2 receptor antagonists – (H₂RAs) – CALIFORNIA H₂

> **C**imetidine (Tagamet)
>
> **A**n androgenic effect like gynecomastia
>
> **L**iver enzymes interaction (CYP450) with cimetidine
>
> **I**ncreases QTc with other QTc drugs, famotidine & cimetidine
>
> **F**amotidine (Pepcid)
>
> **O**
>
> **R**anitidine (Zantac)
>
> **N**izatidine (Axid)
>
> **I**
>
> **A**djust for renal disease

This book has quick summaries and select individual mnemonics for when you have more time to study.

When you went to another country, you snapped a few photos as vacation reminders. These mnemonic snapshots encapsulate detailed content in the same way. When you review, you can revisit the picture rather than reread whole sections.

How do we begin?

We'll start by memorizing the seven pathophysiologic classes in this book, in order. Our mnemonic is G-M-RINCE, as in Grand Mothers RINCE kids' hair, except it's the French r-i-n-c-e instead of the English r-i-n-s-e. These seven pathophysiologic classes provide the foundation to memorize the gastrointestinal (G), musculoskeletal (M), respiratory (R), immune (I), neuro (N), cardio (C), and endocrine (E) systems' medications.

This G-M-RINCE order places drug classes in this book from easiest to hardest to learn. And then, within each drug class section, I answer a pharmacology question with:

1) A unifying, easy-to-review mnemonic flashcard,

2) A quick summary of the most crucial information,

and

3) Select individual mnemonics which read as stories about interesting drug names or properties.

Finally, I'm showing you a question and answer system that you can use for memorization, and with these flashcards, you'll have an excellent head start.

CHAPTER 1: GASTROINTESTINAL MNEMONIC FLASHCARDS

Often, students make flash cards to help them for quizzes and tests. I know there are online flashcards, as well, and this form of mass practice, spaced out over time, is helpful. However, single drug flashcards are inefficient for comprehensive board exams.

Instead, you want to group medications within cognitive short-term memory limits of 3 to 7 items to speed encoding, storage, and retrieval – the three memorization steps. It's challenging to make images from pharmacology. However, you can use words to make pictures in your mind's eye by going down and across, much like a crossword puzzle.

When I first started drawing these flash cards in order on the whiteboard, I had to look at my notes, but over time, I found they stuck in my brain. When drug names remain in order, adding mechanisms and side effects becomes more natural, so feel free to write in new information.

Note: While this book stands alone in audio and written editions, I believe your memory will work better with both. Each mnemonic answers a pharmacology question. Let's begin.

Question 1. Name four antacids and four side-effects or interactions that concern you about antacids.

ACIDIC MEALS

A_____ (antacid)

C_____ (antacid)

I

D_____P_____ (side effect)

I_____C_____ (drug interaction)

C_____ (side effect)

M_____ (antacid)

E

A_____ N_____ (when / how to take antacids)

L_____ (side effect of magnesium hydroxide)

S_____ (antacid)

1. Antacids – ACIDIC MEALS

> **A**luminum hydroxide (Amphojel)
> **C**alcium carbonate (Tums, Pepto Children's)
> **I**
> **D**ecreased phosphate with Al(OH)$_3$, CaCO$_3$, Mg(OH)$_2$
> **I**ons, chelation with fluoroquinolones, levothyroxine, and tetracyclines
> **C**onstipation from Al(OH)$_3$ and CaCO$_3$
>
> **M**agnesium hydroxide (Milk of Magnesia)
> **E**
> **A**s needed (PRN) rather than scheduled
> **L**axative effect of magnesium hydroxide
> **S**odium bicarbonate (in Alka-Seltzer)

Quick Summary

A standard exam question asks: **When should patients take antacids?** Answer: one to three hours after a meal starts producing acid. You use antacids after there is a rise in the acid, much like high tide, except it's the food that triggers the open floodgates of acid in the stomach.

Therefore, our vertical mnemonic is, in all caps, *ACIDIC MEALS*.

A, aluminum hydroxide, brand **Amphojel** alludes to the liquid's viscosity.

C, calcium carbonate, is the active ingredient in **Children's Pepto**. Pepto Bismol for adults has bismuth subsalicylate as its active ingredient, with an aspirin-like ingredient that could cause Reye's syndrome in children.

What's confusing is that **Children's Pepto** is a tablet. **Adult Pepto-Bismol** is a pink liquid. Usually, the children's dosage form is the liquid, and the adult form is the tablet, which can cause parents to think it's okay to use Pepto-Bismol for children. That's wrong. **Do not use salicylates in children.**

On the NCLEX® for nurses, you will likely only see generic names like aluminum hydroxide and calcium carbonate on the board exam. However, many students and patients remember a brand name first and then the generic. So, when you go to your memory for an analgesic, for example, you don't think acetaminophen; instead, you remember Tylenol, the brand name first. I always recommend still trying to learn the brand names because just learning generics might make you test ready, but you'll be clinically unfit. Let's keep moving.

Calcium carbonate's brand name **Tums** goes with the word "tummy" to help you remember it's an antacid. Don't interchange **calcium carbonate (Children's Pepto)** and **bismuth subsalicylate (Adult Pepto-Bismol)**; they have different ingredients.

D, decreased phosphate with Al(OH)$_3$ (aluminum hydroxide) and CaCO$_3$ (calcium carbonate). For the hypophosphatemia some antacids cause, you can think, "aluminum and calcium are pals with phosphate, binding it up and decreasing the phosphate (PO$_4$) level."

I, ions, chelation with fluoroquinolones, levothyroxine, and tetracyclines. The three multivalent cations — aluminum, calcium, and magnesium — are in alphabetical order. *Multivalent* means a charge is higher than one on ionization. Therefore, aluminum, Al+3; calcium, Ca+2; and magnesium, Mg+2, are multivalent. Multivalent cations chelate with certain antibiotics, including fluoroquinolones (ciprofloxacin, moxifloxacin, levofloxacin), levothyroxine, a thyroid replacement medicine, and tetracyclines (doxycycline and minocycline) another antibiotic class. Chelation, the name of this binding, makes these medicines ineffective. To remember this, think of the letters from chelate, c-h-e-l-a-t-e, *c* for ciprofloxacin, *l* for levothyroxine, and *t*, for tetracycline.

C, constipation, reminds us that aluminum hydroxide and calcium carbonate cause constipation.

M, magnesium hydroxide. Its side effect of diarrhea is a therapeutic effect as a laxative, a drug class we'll talk about in a little while. We want to be careful about high magnesium and especially build-up in patients with chronic kidney disease, CKD. Because chronic magnesium use can cause fluid loss, we also watch for electrolyte imbalances. Magnesium hydroxide's **Milk of Magnesia** looks like dairy milk, which can work similarly to an antacid in calming an acidic stomach. Put together the "milky texture" and the diarrhea of lactose intolerance to remember its laxative effect.

A, as needed (PRN) rather than scheduled refers to antacids use just when we need it.

L, laxative effect of magnesium hydroxide, contrasts the constipating effect of aluminum hydroxide.

S, sodium bicarbonate, is easy to remember because metabolic *alka*losis is an increase in serum bicarbonate (HCO_3^-) concentrations from your acid-base and electrolyte study. Sodium bicarbonate, an ingredient in **Alka-Seltzer**, makes a stomach more alkaline, or less acidic, with a bubbling seltzer-like effect.

Why didn't I use all the letters in ACIDIC MEALS? When I asked people to preview my book manuscript, the unfilled letters created a sense of incompleteness. At first, I thought the missing letters were a problem. Then, I read about the Zeigarnik effect from psychology. It says you remember incomplete tasks better than complete ones. This incompleteness may improve the mnemonic's stickiness.

The next step after failing an antacid for chronic heartburn might be the H_2-blockers in the next slide.

GASTROINTESTINAL MNEMONIC FLASHCARDS

Question 2. Name four histamine-2 receptor blockers, their class suffix, two specific concerns about cimetidine, one about cimetidine and famotidine, and a general concern for the drug class itself.

CALIFORNIA H2

C_____ (histamine-2 receptor blocker)

A_____ (concern for cimetidine)

L_____ (concern for cimetidine)

I_____ (concern for cimetidine and famotidine)

F_____ (histamine-2 receptor blocker)

O

R_____ (histamine-2 receptor blocker)

N_____ (histamine-2 receptor blocker)

I

A_____ for R_____ D_____ (general class concern)

2. HISTAMINE-2 RECEPTOR ANTAGONISTS (H₂RAs) – CALIFORNIA H₂

> **C**imetidine (Tagamet)
> **A**ndrogenic effect gynecomastia with cimetidine
> **L**iver enzymes interaction (CYP450) with cimetidine
> **I**ncreases QTc with other QTc, famotidine and cimetidine
> **F**amotidine (Pepcid)
> **O**
> **R**anitidine (Zantac)
> **N**izatidine (Axid)
> **I**
> **A**djust for renal disease

QUICK SUMMARY

We remember the H₂-blockers end in "-tidine, t-i-d-i-n-e." Think of a dining room table where people are going *to dine, t-o d-i-n-e* because, after they've dined, that's when acidic problems like reflux appear. How are H₂-blockers different than antacids? They reduce stomach acid like antacids, but for much longer.

I used *CALIFORNIA* to hold the four histamine-2 receptor antagonists because *cali* means *hot* and *forn* means *oven*. That's how someone with heartburn feels. I related the state of California to H₂ because former California Governor Schwarzenegger had a hydrogen-fueled Hummer H₂ SUV and hydrogen is a big part of acidity. Imagine hearing Arnold Schwarzenegger's voice saying, "California H₂".

GASTROINTESTINAL MNEMONIC FLASHCARDS

C, cime_ti_dine, we don't often see in clinical practice because it has cytochrome P450 interactions, condensed as C-Y-P 450 and pronounced sip, like *sip a drink*. Cimetidine's brand **Tagamet** is an an_ta_gonist.

A, androgenic effect like gynecomastia, is a condition of enlarged male breast tissue that could happen with cimetidine. This rare side effect is a board exam favorite.

L, liver. CYP450 interactions are an issue with inhibiting enzymes. If you impede liver enzymes, then those enzymes can't process other drugs, raising the other drugs' levels in the body.

I, increases QTc with other QTc drugs a concern with famotidine and cimetidine. You can get life-threatening ventricular arrhythmias called Torsade de pointes if the drug interaction prolongs the QT interval resulting in sudden cardiac death.

F, famo_ti_dine's brand **Pepcid** contains "pep" from "peptic" and "pepsin" that relates to digestion and "cid" from "acid." A student mentioned Pepsi, the soda, has a very acidic 2.4 pH. She said it always gave her heartburn, so that's how she remembered **Pepcid**. Also, in soda pop, the "p-o-p" and **Pepcid's** "p-e-p" are very similar.

O,

R, rani_ti_dine 's "-tidine, t-i-d-i-n-e" looks like "to dine," the time patients might experience gastroesophageal reflux disease (GERD). **Zantac** looks like a "2" with "antac" after it, so it's an H₂ blocker, which works as an _ant_-agonist to _ac_-id. Zantac.

19

Memorizing Pharmacology Mnemonics

N, nizatidine's brand **Axid** replaces *acid*'s *c* with an *x*.
Nizatidine rounds out the histamine-2 receptor antagonists. We can use H_2 blockers like these or the proton pump inhibitors that appear next for breakthrough nighttime acid.

I,

A, adjust for renal disease with this group.

Before we go on to our next therapeutic step, moving on to a proton pump inhibitor, PPI, let's take a minute to add a note about side effects and connecting more drugs to them.

Question 3. Name eight drugs and or drug classes that might cause gynecomastia.

FEMALE PARTS CARE

P_____ (HIV drug class)

A_____ (Hormone drug class)

R_____ T_____ (HIV drug class)

S_____ (Diuretic drug)

C_____ (H2RA)

A_____ (hormone drug class)

R_____ (Antipsychotic drug)

E_____ (Hormone drug class)

3. Gynecomastia Side Effect Drugs - Female Parts Care

> **P**rotease inhibitors
>
> **A**ntiandrogens
>
> **R**everse-
>
> **T**ranscriptase inhibitors
>
> **S**pironolactone (Aldactone)
>
> **C**imetidine (Tagamet)
>
> **A**nabolic steroids
>
> **R**isperidone (Risperdal)
>
> **E**strogens

Quick Summary

These drugs can cause gynecomastia, swollen breast tissue in males. Since cimetidine ties to the group of H_2-blockers in the slide before, we can now add to our memory other drugs that also cause gynecomastia. If we don't do this, the mnemonic has nothing to connect to and may fade quickly. Anchors and transitions are critical to keeping knowledge glued to your grey matter.

To get to the next slide, we think of therapeutic progression. After failing an H_2-blocker, one might move on to a proton pump inhibitor, PPI.

GASTROINTESTINAL MNEMONIC FLASHCARDS

Question 4. Name six proton pump inhibitors and their class suffix.

ULCER DEVELOPER

D_____ (PPI right enantiomer)

E_____ (PPI left enantiomer)

V

E

L_____ (PPI)

O_____ (PPI)

P_____ (PPI)

E

R_____ (PPI)

Class suffix

-p_____

23

4. Proton Pump Inhibitors (PPIs) – Ulcer Developer

> **D**exlansoprazole (Dexilant)
> **E**someprazole (Nexium)
> **V**
> **E**
> **L**ansoprazole (Prevacid)
> **O**meprazole (Prilosec)
> **P**antoprazole (Protonix)
> **E**
> **R**abeprazole (AcipHex)

Quick Summary

Often, someone who is an "ulcer developer" needs proton pump inhibitors for treatment. Our acrostic becomes *DEVELOPER*.

Before I go into the proton pump inhibitors, let's pause to talk about mistakes with suffixes. I just read another pharmacology mnemonics book that got this wrong. The World Health Organization, WHO, globally, and the United States Adopted Names Council, USANC, agree on a drug beginning, prefix, middle, infix, or an ending, a suffix, to standardize related drugs.

The ending these organizations agreed on for the H_2-blockers is –tidine, -t-i-d-i-n-e. I see students quoting quizzing notecard websites as reliable sources. They write that drugs ending in –ine, i-n-e, are always H_2-blockers. That's untrue. Morphine, an opioid, and fluoxetine, an antidepressant, both end in –ine, for example. This 'build your own suffix' error is a cognitive heuristic bias. That means the notecard and book authors believed something wrong to be right.

Back to our proton pump inhibitors, PPIs. The ending or suffix is "–prazole, p-r-a-z-o-l-e." The suffix is not "-azole, -a-z-o-l-e," an organic chemistry molecule. If you use that as your mnemonic or suffix, then fluconazole, an antifungal, is grouped into proton pump inhibitors. The proper suffix for fluconazole is "–conazole, c-o-n-a-z-o-l-e" suggesting antifungal. There's also aripiprazole with the "–piprazole, p-i-p-r-a-z-o-l-e" stem for schizophrenia. Using suffixes is powerful, but when an author cites various suspect internet sources, you might want to look elsewhere.

With this –prazole ending, we can now talk about six proton pump inhibitors. Some students connect the *-prazole* stem with *preventing*, and *protons* all starting with the *pr* sound.

To remember this order, we'll use the mnemonic ulcer developer with these drugs, using the *D, E, L, O, P,* and *R* of the word "DEVELOPER."

D, dexlanso<u>prazole</u> brand **Dexilant** stands for *right-handed excellent* because the right-handed enantiomer lasts longer than its racemic lansoprazole counterpart does.

E, esomeprazole. While the "prazole, p-r-a-z-o-l-e" ending means PPI, the "e-s" means "S" for sinister or left-handed, the active enantiomer. The manufacturer released **Nexium** after **Prilosec** as the "next" PPI drug.

V,

E,

L, lansoprazole's **Prevacid** prevents acid.

O, omeprazole's **Prilosec** has the "P-r" for hydrogen protons (protons, or hydrogen ions, are what make an acid acidic), the "lo, l-o" for "low, l-o-w," and the "sec" for secretion of those protons. Or, the "o" in Prilosec looks like a zero, and **Pril-"O"-sec** provides zero heartburn.

P, pantoprazole's brand **Protonix** reverses *nix* and *protons*. Pantoprazole, like esomeprazole and rabeprazole, comes in an IV form.

E,

R, rabeprazole brand **AcipHex** combines *aci* from *acid*, pH (little *p* capital *H*) from the pH scale, and *ex*, to excrete.

We try to avoid working on more than four drug names at a time because it taxes our short-term memory. Here we have six drug names. However, two drug pairs have parallel roots.

Esomeprazole and *omeprazole* only differ by *e-s, es*. These two letters, e-s, mark **esomeprazole** as the left-handed enantiomer, the one with therapeutic benefit. **Omeprazole** is an *es* and *ar* enantiomer mixture with the o-m-e-p-r-a-z-o-l-e root.

The other pair includes **dexlansoprazole**, the (R)-(+)-enantiomer of **lansoprazole**, the racemic mixture with lansoprazole roots.

Your brain chunks these four medications into two couples, like your credit card chunks four groups of four numbers. *Esomeprazole* pairs with *omeprazole* and *dexlansoprazole* pairs with *lansoprazole*. Then it adds two singles, *pantoprazole* and *rabeprazole*, to chunk down to a group of four units like this:

1. **Dexlansoprazole** and **Lansoprazole**
2. **Esomeprazole** and **Omeprazole**
3. **Pantoprazole**
4. **Rabeprazole**

Knowing these groupings is what makes pharmacology easier. If you had a pile of four hundred coins stacked with pennies, nickels, dimes, and quarters, you could separate them by their properties. The pennies are copper. The quarters are largest and dimes smallest. What's left are nickels. In pharmacology, we need to take two minutes to figure out what the properties are for the individual drug classes. Then it's easy to sort them.

Although I focused on peptic ulcer disease, you can use these PPIs for conditions like gastroesophageal reflux disease, GERD, which can (rarely) turn into Barrett's esophagus.

Barrett's esophagus, when tissue like the intestinal lining replaces the esophageal tissue, often comes with a higher risk of esophageal cancer. Regular doctor visits for dysplasia, or precancerous cells, are critical.

Ideally, a patient should take these medications 30 minutes before a meal to get the maximum benefit. Make sure to let the patient know that these medicines may take a few days to take effect and don't work as quickly as antacids.

Now let's take a look at the side effects related to PPIs.

Question 5. Name seven possible side-effects or interactions of proton pump inhibitors.

ULCER DEVELOPER SIDE EFFECTS

Side-effects

D_____ (side effect)

E_____ (side effect)

V

E_____ (interaction)

L_____ M_____ (side effect)

O_____ (side effect)

P_____ (side effect)

E

R_____ H_____ (side effect)

5. Proton pump inhibitors (PPIs) – ULCER DEVELOPER SIDE EFFECTS

> **D**iarrhea from *Clostridium difficile* infection
> **E**lderly, to remind us of dementia
> **V**
> **E**nzymes (CYP issues with some PPIs)
> **L**ow magnesium
> **O**steopenia (bone loss), a concern for fractures
> **P**neumonia
> **E**
> **R**ebound hypersecretion

Quick summary

We use the same letters to tie the side effects to the drugs themselves in this repeat mnemonic.

When the PPI alone doesn't work, you might begin treatment for destroying *Helicobacter pylori*, the active organism in an ulcer.

We'll look at triple and quad therapy next.

Question 6. Name two antibiotics in a triple peptic ulcer disease regimen and how we dose the PPI.

ACED PUD

Drugs for a triple peptic ulcer disease regimen

A_____ (penicillin)

C_____ (macrolide)

E_____ (PPI)

How to dose the PPI in PUD

D_____ D_____

6. Triple Peptic Ulcer Disease (PUD) Therapy – ACED PUD

> **A**moxi<u>cillin</u> (Moxatag)
> **C**lari<u>thromycin</u> (Biaxin)
> **E**some<u>prazole</u> (Nexium)
> **D**ouble dosing of the PPI

Quick Summary

Triple therapy means we are using three drugs, usually one acid reducer and two antibiotics. Quad therapy adds another antimicrobial for four drugs in total. After a recent update to *H. pylori* preferred drug regimens, a round of first-line triple therapy now includes metronidazole, but it would be confusing to throw it in this mnemonic.

Peptic ulcer disease describes ulcers that develop in the stomach or duodenum and extend deep into the mucosa. Ulcers can lead to GI bleeds. Three main peptic ulcer disease causes are stress, NSAID use, and *H. pylori*, a helicopter-like organism that burrows into the stomach lining. The mainstay treatment for *H. pylori* ulcers includes antibiotics to wipe out infection and PPIs to reduce acid. Treatment typically lasts ten to fourteen days.

Traditionally, *H. pylori* infection treatments have included triple therapy – a PPI, amoxicillin, and clarithromycin. For triple therapy, I use *ACED* as the mnemonic as the three-drug combination *ACED* killing *Helicobacter pylori*, the causative organism. You can include any of the PPIs in the treatment regimen, but esomeprazole's *e* made creating mnemonics easier.

A, amoxicillin, a penicillin antibiotic, the -cillin stem lets us know **amoxicillin** is a penicillin antibiotic. With amoxicillin side-effects, we worry about penicillin-allergic patients.

C, clarithromycin, a macrolide antibiotic, Clarithromycin has a "-thro-, t-h-r-o" substem with "–mycin, m-y-c-i-n" as a suffix. Or you could look at the "throw, t-h-r-o" which rhymes with the "cro, c-r-o" in macrolide. Be careful with -mycin; it only lets you know the bacteria comes from the *Streptomyces* class, but it does not tell you the antibiotic class. For clarithromycin, we think of the metallic taste and QT prolongation.

E, esomeprazole, a proton pump inhibitor.

D, double dosing of the PPI means we use BID, twice-daily dosing, rather than traditional QD, once-daily dosing.

Why do we use two or three antibiotics? If you use more than one antimicrobial, you can reduce resistance and decrease dosages to lower the risk of side effects. **Why do we use quad therapy?** If a patient is penicillin allergic, we might use the BED-M regimen which we'll look at next.

Question 7. Name four drugs in each of two quad peptic ulcer disease regimens.

BED-MIDNIGHT OR MACE

Quad peptic ulcer disease regimen #1

B_____ (antidiarrheal)

E_____ (PPI)

D_____ (tetracycline)

M_____ (nitroimidazole)

Quad peptic ulcer disease regimen #2

M_____ (nitroimidazole)

A_____ (penicillin)

C_____ (macrolide)

E_____ (PPI)

7. QUAD PEPTIC ULCER DISEASE THERAPIES - BED-MIDNIGHT OR MACE

> **B**ismuth sub<u>sali</u>cylate (Pepto-Bismol)
> **E**some<u>prazole</u> (Nexium)
> **D**oxy<u>cycline</u> (Vibramycin)
> **M**etro<u>nidazole</u> (Flagyl)
>
> **M**etro<u>nidazole</u> (Flagyl)
> **A**moxi<u>cillin</u> (Moxatag)
> **C**lari<u>thromycin</u> (Biaxin)
> **E**some<u>prazole</u> (Nexium)

Quick Summary

For quad therapy, I use *BED-M* (think: "acid in bed at midnight") and MACE, a weapon, crushing *H. Pylori*.

B, bismuth, has antibiotic properties. Bismuth can have adverse side effects, such as Reye's syndrome, a neurologic condition. Also, watch for salicylism, and tarry black stools and black tongue. The black stools may mimic symptoms of GI bleed, so that's a counseling consideration.

E, esomeprazole, is a proton pump inhibitor to reduce acid.

D, doxycycline, a tetracycline antibiotic. Like using esomeprazole as representative for any PPI, we can substitute other tetracyclines. Doxycycline is a tetracycline we avoid with children to prevent tooth discoloration. Watch also for chelation and sun sensitivity.

M, metronidazole, an antiprotozoal that can cause metallic taste and nausea. If a patient adds alcohol, this causes a disulfiram reaction that may result in severe vomiting. The MACE mnemonic reorders drugs we've talked about before.

M, metronidazole

A, amoxicillin

C, clarithromycin

E, esomeprazole

Our next slide uses metronidazole as an anchor drug to connect to the disulfiram reaction of alcohol and another list of medications.

Question 8. Name six medications known to cause the disulfiram reaction and the reaction's underlying mechanism.

MEGA PUCED

Disulfiram reaction medications

Me_____ (nitroimidazole antibiotic)

G_____ (antifungal)

A_____ of A_____ (underlying mechanism)

P_____ (chemotherapy)

U

Ce_____ and Ce_____
(cephalosporin antibiotics)

D_____ (for alcoholism)

8. Disulfiram Reaction – MEGA PUCED (PUKED)

> $M\text{e}$tronidazole (Flagyl)
>
> Griseofulvin (GrifulvinV)
>
> Accumulation of acetaldehyde, a toxic metabolite
>
> Procarbazine (Matulane)
>
> U
>
> $C\text{e}$fotetan (Cefotan, 2nd) and $C\text{e}$foperazone (Cefobid, 3rd)
>
> Disulfiram (Antabuse)

Quick Summary

The disulfiram reaction is intense nausea and vomiting that comes from mixing alcohol and medication. Metronidazole serves as the anchor drug to connect to a previous slide. The mnemonic MEGA PUCED, P-U-C-E-D using the hard "C" sound for PUKED, P-U-K-E-D. The mnemonic reminds us that if someone mixes these medications with alcohol, we'll see severe vomiting.

ME, metronidazole is an antiprotozoal.

G, griseofulvin is an antifungal.

A, accumulation of acetaldehyde, a toxic metabolite represents the mechanism of action of the reaction.

P, procarbazine is for Hodgkin and non-Hodgkin lymphoma.

U, is a letter placeholder.

CE, <u>Ce</u>fotetan and <u>Ce</u>foperazone are cephalosporin antibiotics.

D, disulfiram is for preventing alcohol abuse. The brand name Antabuse is Anti-abuse minus the letter 'i' and the hyphen.

Let's move to the next slide, which includes medications that protect the stomach from damage and either prevent ulcers or allow them to heal.

Memorizing Pharmacology Mnemonics

Question 9. Name a prostaglandin-like drug to protect the stomach, two prokinetics, and a protectant.

MIMES STUCK IN STOMACH

Mi_____ (prostaglandin-like drug)

M_____ (prokinetic)

E_____ (prokinetic, antibiotic)

S_____ (protectant)

9. Prostaglandin, two prokinetics, and a protectant – MIMES STUCK IN STOMACH

> ***Mi**soprostol* (Cytotec)
> ***M**etoclopramide* (Reglan)
> ***E**rythromycin* (E-Mycin)
> ***S**ucralfate* (Carafate)

Quick Summary

Let's pretend, however, that we see an ulcer in our stomach, and there are no *Helicobacter pylori*. What happened? It's probably a non-steroidal anti-inflammatory drug, an NSAID induced ulcer. The NSAIDs include ibuprofen, naproxen, and meloxicam.

A drug to prevent NSAID induced ulcers is **misoprostol**, brand **Cytotec**, but this is controversial. The fetal harm is so high practitioners use it for abortions.

Another thought is to use the prokinetic **metoclopramide**, brand **Reglan** or the antibiotic **erythromycin** Acid leaving the stomach faster results in less ulcer injury. Think of opening the acidic faucet from our stomach.

On the other hand, we might want to coat the bottom of the stomach to protect that ulcer with **sucralfate**, brand **Carafate**. This pink liquid covers the stomach and provides a barrier until the ulcer can heal. We hold these drug names in our minds where they work, at the bottom of the stomach.

One of my students better sees concepts connect when she puts them on the board. As I wrote the drug classes for these medications, I noticed they all began with *p-r-o*: prostaglandin, prokinetic, and protectant. For the acrostic, *MIMES*, I took the *m-i* from misoprostol, the *m* from metoclopramide, *e* from erythromycin and the *s* from sucralfate.

When two drugs begin with the same letter, like misoprostol and metoclopramide, use more than one letter to build at least one of the mnemonics. The MIMES STUCK IN STOMACH mnemonic helps us picture a mime in his invisible wall trying to get out. The pro KIN etics, part of stucK IN, help him leave."

MI, misoprostol, is a prostaglandin analog, hence the *prost* in the name, typically used for NSAID-induced ulcers. NSAIDs inhibit prostaglandins, so we use misoprostol to replace them. It's a prostaglandin E1 analog, which works on GI tract parietal cells. Use the *m-i-s* to remember that misoprostol has some miserable side effects like stomach cramping and diarrhea, so that adherence can be an issue. Misoprostol's brand **Cytotec** takes some letters from adenylate cyclase, which leads to decreased proton pump action, but that isn't as helpful as the generic misoprostol's *prost* evoking prostaglandin.

M, metoclopramide, is a promotility agent. Delayed gastric emptying can cause GERD, and metoclopramide can encourage gastric emptying. Remember the mechanism of action by thinking about *clo* in metoclopramide rhyming with *flow*, which suggests moving food from the stomach to the intestines. Metoclopramide is an anticholinergic agent, so think of drying symptoms when you think of the side effects (dry mouth, constipation, blurry vision). Additionally, we might see gynecomastia, an increase in male breast tissue size. Watch for hyperprolactinemia, elevated serum prolactin that results in oligomenorrhea (irregular menstrual periods), amenorrhea (no menstrual period), galactorrhea (milky discharge from the breasts), or hirsutism, excessive facial hair growth.

We also see metoclopramide for nausea and vomiting. If you want to get specific, metoclopramide blocks dopamine, which would typically cause nausea and vomiting in the medullary chemoreceptor trigger zone, CTZ. If we affect dopamine, we have to watch out for movement disorders like extrapyramidal symptoms. Parkinson's patients, with an early movement disorder, make this a troublesome choice. Metoclopramide's **Reglan** is a dopamine receptor antagonist taking the *re* from *receptor* and the *"an"* from *antagonist*.

E, erythromycin, an antibiotic so known for its GI effects that prescribers employ it as a prokinetic.

S, sucralfate, is a mucosal protectant that prevents damage to the stomach mucosa. It's like when, after you drink a glass of milk, you can still see the milky-white coating around the glass; that's what sucralfate does to your stomach. It's usually used to protect the stomach after a patient has an ulcer. It does not affect acid production. We can remember its mechanism by thinking, "the fate of your stomach after an ulcer depends on sucralfate." Sucralfate is an aluminum salt with similar side effects to aluminum hydroxide like hypophosphatemia and a need for renal dosing. Sucralfate's brand name **Carafate** reminds me of the word *crater*, something that it protects from acid or a carafe, like a carafe of acid.

We've covered many drugs that protect the stomach, so let's move down to drugs that affect the intestines, pairing medications for:

DC, diarrhea and constipation,

NV, nausea and vomiting,

IBS, IBD, irritable bowel S, syndrome and inflammatory bowel D, disease

The two-letter abbreviations, DC, NV, and SD from the end of IBS, and IBD, immediately brought to my mind Washington D.C., Nevada, and South Dakota. That's how I remember which order to use when going over these medication classes as a teacher.

Question 10. In order of aggressiveness of therapy from least to most, name three antidiarrheal or combination drugs.

BUILD A BARRIER

B_____ (salicylate, least aggressive)

U

I

L_____ (opioid receptor agonist)

D_____/

A_____ (opioid, anticholinergic combination, most aggressive)

BARRIER

10. HAVE DIARRHEA? – BUILD A BARRIER

> **B**ismuth sub<u>sa</u>licylate (Pepto-Bismol)
> *U*
> *I*
> **L**operamide (Imodium)
> **D**iphenoxylate /
>
> **A**<u>tropine</u> (Lomotil)

QUICK SUMMARY

Diarrhea can lead to dehydration, and sometimes we need to intervene and use over-the-counter medications like **bismuth sub<u>sa</u>licylate** or **loperamide**. If it is severe, we might add the controlled substance **diphenoxylate**.

Diarrhea and constipation go together as opposites. To tie them to something familiar, I thought of Washington D.C., just to put the letters *D* and *C* together. I used the mnemonic BUILD for three antidiarrheals. Think of "build a barrier" to fecal flow. From least to most aggressive treatment we have:

B, bismuth sub*sa*licylate, is an OTC drug. Bismuth's *subsalicylate* is similar to aspirin (acetyl<u>salicy</u>lic acid) and is dangerous to young children. **Bismuth sub<u>sa</u>licylate** is not for children because of Reye's syndrome risk, a condition involving brain and liver damage that can occur in children with chickenpox or influenza who take **salicylates**. The "b" in **bismuth subsalicylate** reminds students of the black tongue and black stool that some patients experience as side effects. (Note: This discoloration is harmless.) **Pepto** looks like peptic, which has to do with digestion.

L, loperamide, a slightly more aggressive therapy, acts on opioid receptors in the intestines, so we don't worry about addictive potential. With loperamide, often we miss what's most important, that diarrhea is a symptom, not a disease. Because this is readily available over-the-counter, patients may self-treat conditions like medication side effects or irritable bowel syndrome, without fixing the underlying problem. *Lo* in s<u>lo</u>w and *per* in <u>per</u>istalsis is how you remember **loperamide's** function. **Imodium** is like the word *immobile* –, in that it slows the bowel.

D, diphenoxylate with <u>atropine</u>, is a prescription-strength, DEA Schedule V medication. **Atropine** prevents someone from crushing **diphenoxylate** and injecting it illicitly. The brand name **Lomotil** suggests "<u>lo</u>w <u>motil</u>ity" – since it slows diarrhea.

Side note: **Octreotide**, absent from the mnemonic, is for refractory irinotecan-induced diarrhea, a specialized treatment.

Next, we'll take a look at anticholinergic side-effects like those that atropine causes.

Memorizing Pharmacology Mnemonics

Question 11. Name the six anticholinergic side-effects that you might expect with a drug like atropine.

ABDUCT WATER

A_____

B_____ V_____

D_____ M_____

U_____ R_____

C_____

T_____

11. ANTICHOLINERGIC SIDE EFFECTS – ABDUCT WATER

> **A**nhidrosis, lack of sweating
> **B**lurry vision, secondary to dry eyes
> **D**ry mouth or xerostomia
> **U**rinary retention
> **C**onstipation
> **T**achycardia

QUICK SUMMARY

Diphenoxylate with atropine is an opioid and anticholinergic combination. In the word 'anticholinergic,' anti means against and cholinergic implies acetylcholine – so, against the neurotransmitter acetylcholine.

A synonym is antimuscarinic referencing muscarine, a natural product found in some mushrooms. The anticholinergic mnemonic for side effects that I use is *ABDUCT*, as in *ABDUCT WATER*.

Those six anticholinergic effects – anhidrosis, blurry vision, dry mouth, urinary retention, constipation, and tachycardia – are the opposite of the SLUDGE, S-L-U-D-G-E mnemonic you may have heard for a cholinergic crisis. Cholinergic is the opposite of anticholinergic, just as muscarinic is the opposite of antimuscarinic.

We now continue with *cholinergic* side-effects.

Memorizing Pharmacology Mnemonics

Question 12. Name six cholinergic side-effects.

SLUDGE WATER

S_____

L_____

U_____ I_____

D_____

G_____ D_____

E_____

12. CHOLINERGIC SIDE EFFECTS – SLUDGE WATER

> **S**alivation, opposite dry mouth
> **L**acrimation, opposite blurry vision with dry eyes
> **U**rination incontinence, opposite urinary retention
> **D**iarrhea, opposite constipation
> **G**astric distress and
> **E**mesis

QUICK SUMMARY

Instead of using anticholinergic and cholinergic in the classroom, we use them as adjectives. A student asks, "is it an ABDUCT drug (anticholinergic) or a SLUDGE drug (cholinergic)?" Representing acetylcholine's effects with single-syllable mnemonics simplifies talking about them.

From diarrhea treatments, we move to constipation medicines.

Memorizing Pharmacology Mnemonics

Question 13. In order of least to most aggressive treatment, name four drugs for constipation.

PRE-DISPOSED

P_____ (fiber)

R

E

D_____ (stool softener)

I

S_____ (stimulant)

PO_____ G_____ (osmotic)

S

E

DO

52

13. HAVE CONSTIPATION? – PRE-DISPOSED

> **P**syllium (Metamucil)
> **R**
> **E**
> **D**ocusate sodium (Colace)
> **I**
> **S**enna (Senokot)
> **Po**lyethylene glycol (GoLytely, MiraLAX)
> **S**
> **E**
> **D**

QUICK SUMMARY

Constipation comes from diet, medications, or health conditions. I ordered them by container size, stacked from left to right, as well as the order of laxative intensity. For a mnemonic, think of fecal matter PRE-DISPOSED; before it's dumped.

P, psyllium, comes in a sizeable towering container and represents the least invasive option – adding fiber to a person's diet. Psyllium is fiber that can add bulk to stools but takes up to three days to start working. It's the safest of the four choices. Make sure to include plenty of water when giving the medicine, as psyllium can cause gas and bloating. Psyllium is dietary fiber. The *psylli* part of psyllium, which we pronounce like *silly*, roughly rhymes with dietary. Also, you can find the letters *m-i-l-l* in *psyllium*, which reminds me of a fiber mill.

D, docusate sodium, stands a bit shorter and softens the stool rather than causing a laxative effect. Docusate sodium is less a laxative than it is a medicine meant to prevent constipation, especially when caused by opioids like hydrocodone or oxycodone. Docusate pulls water into the intestines to soften up the stool. Docusate and *penetrate* rhyme, and docusate sodium works by helping water move into the bowel. The brand **Colace** improves the colon's pace.

S, senna, is in a smaller bottle still and is a stimulant laxative for when you get all mush and no push. The *senna* and *stimulants'* "s" tie them together. It usually works in 6 to 12 hours for faster relief. A related drug, **bisacodyl** (**Dulcolax**) comes as a suppository (and other forms) if the patient struggles to swallow. Senna sends everything into the toilet. You may see the brand **Senokot-S**, which is a combination medication that contains senna and docusate. I expect the letter S stands for an *S, softener*.

PO, polyethylene glycol (PEG), the small purple measuring capful worth of **MiraLAX**, anchors the group. MiraLAX is a miracle laxative because it's a miracle how good you feel after taking it. There is another version of PEG that provides bowel prep – the brand **GoLytely**, a gallon jug of water and powder for colonoscopy examination preparation works quickly, and *lightly* doesn't describe what happens next; it has a powerful laxative effect.

Here are some quick notes on therapeutics. Opioids like **morphine** decrease gastrointestinal tract motility, causing constipation. Calcium channel blockers like **verapamil** block calcium from getting to the bowel's smooth muscle. Prescribers often give a stool softener like **docusate sodium** and a stimulant laxative like **senna** for constipation when prescribing these constipation-causing medications to help patients stay ahead of and prevent the constipating effect.

With an osmotic laxative like **polyethylene glycol**, we might run into its relatives on an exam. One of these is **lactulose (Enulose)**, which helps reduce levels of ammonia in hepatic encephalopathy or neuropsychiatric issues in patients with liver disease like cirrhosis. For this kind of treatment, we'll see much higher doses. Often, a patient will get a grocery bag of multiple pint-sized bottles to take home.

Another constipation choice is the **glycerin suppository**, often for young patients.

Next, let's take a look at medications related to nausea.

Question 14. Name an antivertigo drug for cruise ship travel. Name two drugs you would use to prevent **CINV**, one drug that you would add on if the first two don't work alone, and a PO and IV drug for delayed emesis.

STOP ADD AFTER

S_____ (for cruise ship travel)

T

O_____ (to prevent CINV)

P_____ (to prevent CINV)

A

D

D_____ (add-on drug to prevent CINV)

A_____ (PO drug for delayed emesis)

F_____ (IV drug for delayed emesis)

T

E

R

14. HAVE NAUSEA? – STOP ADD AFTER

> **S**copolamine (Transderm-Scop)
> **T**
> **O**ndan<u>setron</u> (Zofran)
> **P**alono<u>setron</u> (Aloxi)
>
> **A**
> **D**
> **D**examethasone (Decadron)
>
> **A**pre<u>pitant</u> (Emend)
> **F**osapre<u>pitant</u> (Emend injectable)
> **TER**

QUICK SUMMARY

I set these six medications in a specific order based on how they work and what dosage forms you use. The STOP mnemonic reminds you that these drugs stop nausea.

Motion Sickness

S, scopolamine, which works on nausea induced by motion sickness. It's an anticholinergic applied behind the ear at least four hours before the patient needs it. Scopolamine works for 72 hours. Side effects include dry mouth, blurred vision, drowsiness, and constipation. **Transderm-Scop** is a transdermal form of scopolamine for motion sickness. "Trans" means across, "derm" means "skin," so across-the-skin gives us scopolamine – Transderm-Scop.

CINV (Chemotherapy-induced nausea and vomiting)

This is a very different type of reason for being nauseous - chemotherapy treatments. For those, we begin with the "-setrons," which are serotonin 5HT$_3$ drugs ending in –s-e-t-r-o-n, ondansetron, and palonosetron. When I find two drugs that have different dosage forms, I've always put the PO, or oral medicine, first, and then put the IV, intravenous dosage form, second.

These 5HT$_3$s are especially useful for **c**hemotherapy-**i**nduced **n**ausea and **v**omiting, CINV. Ideally, if we can stop the brain from wanting to vomit, we can help patients on chemo who are already weak from medication.

We classify CINV as acute, within 24 hours of chemotherapy, or delayed, more than 24 hours after. Emetogenecity, or how likely a chemotherapeutic drug causes nausea and vomiting determine which anti-nausea drugs to give before chemotherapy. Patients receiving highly emetogenic cisplatin and cyclophosphamide often need an NK$_1$ antagonist, dexamethasone, and a 5-HT$_3$ antagonist before chemo. Patients with moderately emetogenic agents get dexamethasone and a 5-HT$_3$ antagonist. There are many single-therapy options for patients with low emesis risk.

O, ondansetron is an O, oral, 5HT$_3$ serotonin receptor antagonist. The *–setron* suffix will help you remember **ondansetron** is a serotonin 5-HT$_3$ receptor antagonist for preventing emesis. Ondansetron *ODT* is short for *an orally disintegrating tablet*. It is a useful dosage form because it dissolves on the top of the tongue and needs no added liquid. If you are good at word scrambles, *ondansetron* contains every letter, except *i*, in *serotonin*, a neurotransmitter in the GI tract. Lots of stomach serotonin can trigger nausea/vomiting, and ondansetron blocks this.

P, palonosetron, is the P, parenteral form with a longer half-life. Notice the *l-o-n* in **palonosetron** to remind you of its long half-life. We associate high emetogenicity with lots of serotonin, so an antagonist fits here. Watch for headaches and QT issues at the higher doses. Like its sister drug ondansetron, palonosetron is a 5-HT$_3$ receptor antagonist. Palonosetron is available IV, useful for V, vomiting patients. *Pal* in *palonosetron* is similar to a *pail*, dumped straight into the vein. Palonosetron has a long half-life, which works for delayed CINV.

When you see serotonin as a drug class, you also want to watch for other drugs affecting the same neurotransmitter as selective serotonin reuptake inhibitors, SSRIs. Also, you may see the QT prolongation risk with ondansetron as a drug interaction question. If ondansetron doesn't work alone, we add a corticosteroid.

D, dexamethasone. Sometimes a provider will add dexamethasone, a steroid, for more robust cases. Think of the long *i* sound in *fight* and *flight* to remember irritability, insomnia, and increased blood sugar, the letter *i* trio of side effects.

Usually, I would put steroids in the respiratory section because of their usefulness with inflammation. However, we see that this progression of aggressive anti-nausea therapy fits well.

While the "-sone, s-o-n-e" in dexamethasone isn't a proper stem, it's still an ending you'll often see in steroids and a helpful way to identify them.

Neurokinin-1 (NK₁) receptor antagonists

Aprepitant** and **fosapre**pitant** are neurokinin 1 (NK₁) receptor antagonists with the substance P effect in the brain. Substance P is the main culprit causing both acute and delayed CINV, and NK₁ receptor antagonists prevent CINV. The NK₁ antagonists are only for CINV, have CYP interactions, and cause GI upset.

I put them in the same order by dosage form that I did with ondansetron and palonosetron – first the enteral or PO form, aprepitant, and second the enteral form, fosaprepitant.

Note the *–pitant* stem. While the "–tant, -t-a-n-t" suffix suggests a tachykinin (neurokinin) receptor antagonist, the "–pitant, p-i-t-a-n-t" refers more directly to an NK₁ antagonist.

I used the *AFTER* mnemonic part to remind us that we use the NK₁ inhibitors for delayed emesis.

A, aprepitant, brand, **Emend**, is available orally. On the box of Emend, you'll see the *Em, e-m* and *end, e-n-d* in differently colored letters. Emend combines the *em* (emesis) and *end*, the needed drug outcome. Emesis end.

F, fosaprepitant, brand **Emend Injection**, is an NK₁ receptor antagonist in IV form.

Side notes:

While we won't see the drug **meclizine** until the respiratory chapter because it's an antihistamine, its primary role is not allergic rhinitis, rather nausea from motion sickness or vertigo. Meclizine is another over-the-counter medication that a person might use to self-treat. Dizziness might come from other medications or a high antihypertensive dose. Check the profile when patients bring this OTC to the counter. You can see anti-vertigo in meclizine's brand name **Antivert**.

For cancer induced nausea and vomiting, **olanzapine** is effective. Recent guidelines recommend it for chemo regimens with high emetic potential.

Lorazepam can prevent anticipatory nausea when a patient had nausea from a prior chemo regimen, and they feel nauseous before the next chemo round.

Granisetron and **dolasetron** also share the -setron ending.

Next, we'll take a look at two drugs for nausea and vomiting.

Question 15. Name two drugs that come in suppository form to prevent vomiting.

PROVIDE MAGAZINE

1. PRO_____AZINE
2. PRO_____AZINE

15. VOMITING? – PROPVIDE MAGAZINE

Prochlorper**azine** (Compazine)
Prometh**azine** (Phenergan)

QUICK SUMMARY

For vomiting, I used the mnemonic *PROVIDE MAGAZINE* because the drug names both begin with *pro* and end with *azine*. Both have suppository forms so that we can treat an emetic patient. You can picture a patient reading between bouts of vomiting as a way to stay occupied while remaining near the toilet.

Vomiting patients often can't take oral medications. We might need the rectal **promethazine** or **prochlorperazine** forms. With promethazine, watch you do not give it via IV push. There's a black box warning that you must dilute it to avoid tissue necrosis. With prochlorperazine, there's the potential for extrapyramidal symptoms because of its effect on dopamine. Treat this with diphenhydramine.

Prochlorperazine and **promethazine** share the same first three letters (*pro*) and last five letters (*azine*). While this isn't a proper stem, both are phenothi<u>azine</u>s, a-z-i-n-e-s. Both happen to be available in suppository forms, as well. The letter "V" in vomiting looks like a rectal suppository.

Promethazine is technically an antihistamine, and manufacturers sometimes combine its liquid form with codeine. Promethazine also reduces nausea. In addition to oral, IM, and

IV forms, **promethazine** comes in a rectal suppository if a patient can't take anything by mouth (PO).

In our next slide, we'll take a look at inflammatory bowel disease, IBD and irritable bowel syndrome, IBS.

Question 16. Name three drugs for inflammatory bowel disease and two drugs for irritable bowel syndrome.

GI MISLED

M _____ (IBD)

I _____ (IBD)

S _____ (IBD)

L _____ (IBS)

E

D _____ (IBS)

16. IRRITABLE BOWEL DISEASE AND IRRITABLE BOWEL SYNDROME – GI MISLED

> **M**esalamine (Pentasa)
>
> **I**nfliximab (Remicade)
>
> **S**ulfasalazine (Azulfidine)
>
> **L**ubiprostone (Amitiza)
>
> **E**
>
> **D**icyclomine (Bentyl)

QUICK SUMMARY

I began with the medications for the more serious irritable bowel disease, mesalamine, infliximab, and sulfasalazine. I followed with lubiprostone and dicyclomine for irritable bowel syndrome. This back and forth diarrhea and constipation lead us to the GI MISLED mnemonic.

IBD, inflammatory bowel disease, is a disorder that causes chronic inflammation of the bowel. IBD includes ulcerative colitis (UC) and Crohn's disease, which cause symptoms like ulcerations and inflammation (-itis) in the colon. Autoimmune diseases occur when the body's immune system inappropriately attacks an area or areas of the body.

UC and Crohn's disease differ in their location in the GI tract and depth of bowel wall inflammation. UC is the less severe form, only affecting the superficial lining. Inflammatory patches are continuous, and just located in the colon and rectum. Crohn's disease inflammation can impact the entire bowel depth and present as patches throughout, making it difficult to treat.

For inflammatory bowel disease, we can use:

M, Mes<u>a</u>lamine. It's brand **Pentasa** alludes to the 5-Aminosalicylic acid, taking the Greek prefix for five, *pent* (think of the five-sided Pentagon in Washington D.C.), and the aspirin abbreviation for aspirin, ASA, to make *Pentasa*.

I, infl<u>iximab</u> blocks the tumor necrosis factor, alpha (TNF-alpha), to treat this disease. Infliximab is a biologic, a genetically engineered protein. Break **infliximab's** generic name as *inf + li + xi + mab*. The *inf* prefix separates similar drugs. The *li* stands for <u>i</u>mmunomodu<u>l</u>ator, the target. The *xi* stands for chimeric (the source; e.g., combining genetic material from a mouse with a human's). The *x* might refer to the Greek letter *chi*, which looks like an *x*. The *mab* stands for <u>m</u>onoclonal <u>a</u>nti<u>b</u>ody. There are other TNF-alpha drugs common to IBD like **adalimumab** with similar interior stems.

Conditions like ulcerative colitis can go into remission. Take r-e-m from remission and a-d-e from aide, **Remicade**, infliximab's brand name, is a <u>rem</u>ission <u>aide</u>.

S, <u>S</u>ulf<u>a</u>s<u>a</u>l<u>a</u>zine takes *sulfa* from sulfanilamide and *sal* from <u>s</u>alicylic acid. You will see the *sal* stem in the aminosalicylates me<u>sa</u>lamine and sulfa<u>sa</u>lazine. These are usually better for ulcerative colitis than Crohn's alone. The *sulfa-* in sulfasalazine is a potential problem, making mesalamine preferred with its various dosage forms.

I wanted a biologic here to remind you that IBD is an autoimmune disorder and often more severe than any IBS condition.

IBS, irritable bowel syndrome, is an abdominal pain with diarrhea, constipation, or both, without an exact cause. Dietary modifications might help IBS.

L, lubiprostone, is a laxative for IBS and chronic idiopathic constipation or constipation where we don't know the cause. Whether it's IBS-associated constipation in women or opioid-induced constipation, it seems helpful. While the -*prost*- stem suggests prostaglandin; it doesn't help with therapeutic recognition. I think, "Use lube to propel a stone from the body." Lubiprostone treats IBS patients experiencing constipation.

E,

D, dicyclomine, helps intestinal spasm symptomology found in irritable bowel syndrome (IBS) and inflammatory bowel disease (IBD). As with all anticholinergics, we try to avoid use in elderly patients. With dicyclomine's brand **Bentyl**, picture a patient "bent" and "ill" with abdominal pain.

Side notes:

Linaclotide (Linzess) is for IBS patients with constipation, and it increases intestinal fluid volume and slows GI transit time.

Plecanatide (Trulance) is a 2017 addition for IBS with constipation. It works like linaclotide, but patients can take it irrespective of food and may cause less diarrhea than linaclotide. However, there is a contraindication in those under 6-years-old because of dehydration and death risk.

Hydrocortisone IV is for severe, refractory cases of Crohn's Disease

Next, we'll move to the musculoskeletal section.

CHAPTER 2: MUSCULOSKELETAL MNEMONIC FLASHCARDS

Memorizing Pharmacology Mnemonics

Question 17. Name six nonsteroidal anti-inflammatory drugs, NSAIDs.

NSAIDS

N_____

S_____

A_____

I_____

D_____

S_____

17. Non-Steroidal Anti-Inflammatory Drugs - NSAIDS

> **N**aproxen (Aleve, Naprosyn)
>
> **S**ulind<u>ac</u> (Clinoril)
>
> **A**spirin (Ecotrin)
>
> **I**bu<u>profen</u> (Motrin, Advil)
>
> **D**iclofen<u>ac</u> (Voltaren)
>
> **S**al<u>sal</u>ate (Disalcid)

Quick Summary

Non-steroidal anti-inflammatory drugs, NSAIDs, relieve pain and inflammation. Many NSAIDs come over-the-counter and are one of the first drugs a patient reaches for, so I thought I'd start here. The NSAIDS mnemonic fits well if you want to remember common drug names, but there is no order to it.

So, I changed it a little bit in the next slide to contain more NSAIDs and helpful information.

Memorizing Pharmacology Mnemonics

Question 18. Name the hormone-like compound NSAIDs affect, the three OTC NSAIDs in order of half-life, nine prescription-only NSAIDs, which NSAID is COX-2 specific, and which NSAID comes as an injectable.

PAINN MEDS KICK IN

P_____ (hormone-like compound)

A_____ (over-the-counter NSAID)

I_____ (over-the-counter NSAID)

N_____ (over-the-counter NSAID)

N_____

M_____

E_____

D_____

S_____

K_____

I_____

C_____ (COX-2 specific)

K_____ (injectable)

In_____ (dosage form)

18. NSAIDs – PAINN MEDS KICK IN

> **P**rostaglandins
>
> **A**spirin (Ecotrin)
>
> **I**bu<u>profen</u> (Motrin)
>
> **N**aproxen (Aleve, Naprosyn)
>
> **N**abumetone (Relafen)
>
> **M**elox<u>icam</u> (Mobic)
>
> **E**todol<u>ac</u> (Lodine)
>
> **D**iclofen<u>ac</u> (Voltaren)
>
> **S**ulind<u>ac</u> (Clinoril)
>
> **K**eto<u>profen</u> (Orudis)
>
> **I**nd<u>omethacin</u> (Indocin)
>
> **C**ele<u>coxib</u> (Celebrex)
>
> **K**etorol<u>ac</u> (Toradol)
>
> **In**jectable form is ketorol<u>ac</u>

QUICK SUMMARY

In this slide, we'll work to remember 12 NSAIDs. To do this, we divide the material into three mnemonics: **PAINN** spelled with two Ns; **MEDS**, short for medicines; and **KICK**, like 'kick the ball;' as well as "I-N, **IN**" to: 1) make it a sentence, and 2) remind us of ketorolac's injectable form.

Mnemonic Part 1. PAINN.

P, prostaglandin, is a hormone-like compound responsible for blood clotting, labor induction, and contracting and relaxing muscles and blood vessels. Prostaglandins release in response to inflammation to cause pain and fever. NSAIDs reverse this and help with arthritis, gout, and headache. Think "A-G-H, AGH," as in "AGH," my…

A, arthritis,

G, gout, and

H, headache hurt — pain.

The next four drugs represented by A-I-N-N, AINN stands for NSAIDs in order of dosing frequency.

A, aspirin's acronym, "A-S-A," comes from the chemical name: Acetylsalicylic Acid, takes "A" from acetyl, "S" from salicylic, and "A" from acid. The brand **Ecotrin** is enteric-coated aspirin, taking the "e" from "enteric," the c-o and t from "coated," and the r-i-n from "aspirin" to make **Ecotrin**.

I, ibuprofen, and **naproxen** both end with "e-n," and that's a good way to remember they are two NSAIDs. However, many drugs end with "e-n." That trick won't help in a comprehensive multiple-choice exam. A better mnemonic tool is to notice that "profen, p-r-o-f-e-n" from **ibuprofen** is a recognized stem and differs from "proxen, p-r-o-x-e-n" in **naproxen** by only one letter.

N, naproxen, has no formal stem, but use this mnemonic for the brand name: **Aleve** will <u>alle</u>viate pain from strains and sprains.

N, nabumetone's brand **Relafen** reminds you of pain 'relief.'

You take aspirin and ibuprofen four times daily, q.i.d., and naproxen and nabumetone twice daily, b.i.d. These first three drugs are also over-the-counter. Nabumetone is by prescription.

We typically use aspirin in low doses for cardiovascular protection, and much less so for pain, fever, and inflammation. Low-dose aspirin was "baby aspirin," but marketing is getting away from this because of the salicylate-induced Reye's syndrome we spoke about in Chapter 1 with bismuth subsalicylate.

Aspirin's mechanism of action differs slightly from other NSAIDs. It also irreversibly inhibits platelet aggregation, so patients should stop aspirin 7-10 days before surgery to prevent bleeding.

NSAIDs commonly cause gastrointestinal side effects like upset stomach, and sometimes GI bleeds and ulcers. Cardiovascular overwork may come from sodium and fluid retention. Finally, watch for renal problems with NSAIDs, especially around renal disease, they are nephrotoxic.

Mnemonic Part 2. MEDS. The four drugs we'll cover here have either the NSAID suffix "i-c-a-m, -icam" or "a-c, -ac." They work like other NSAIDs with similar side effects.

M, melo<u>xicam</u>'s brand **Mobic** can remind us that this NSAID treats 'big' swelling.

E, etodolac, is the first of four drugs with the –ac, a-c stem.

D, diclofenac, which is available orally and as a topical gel. I have a quick story. In Ireland, we could get the gel with a pharmacist conversation and 10 Euros. We can't do this in the U.S. without a physician visit. Why does this matter? If a patient suffering from pain can avoid a systemic drug like ibuprofen or naproxen, they can avoid GI distress.

S, sulindac

Mnemonic Part 3. KICK IN.

K, ketoprofen, shares a suffix with **ibuprofen.**

I, indomethacin, has a high risk of NSAID induced ulcers at high doses. Indomethacin's manufacturers removed the middle "metha" to get the brand name **Indocin**. Think "ind, i-n-d" for high rate of induced ulcers.

C, celecoxib, is an NSAID, but it's more selective for cyclooxygenase-2, COX-2 at therapeutic concentrations and should cause fewer GI effects like upset stomach and bleeding. An oversimplification is that COX-1 equals stomach, COX-2 equals pain, so you only work on COX-2. Celecoxib may, unfortunately, have increased cardiovascular risk, like stroke and heart attack, and this is a black box warning.

Celecoxib's "–coxib" suffix lets you know celecoxib is a selective COX-2 inhibitor. The commercials for **Celebrex** talk about celebrating relief from inflammatory conditions. Regular NSAIDs like **ibuprofen** and **naproxen** inhibit both COX-1 and COX-2, causing the stomach distress so commonly caused by drugs in the NSAID class. **Celebrex** causes less GI irritation due to its lack of COX-1 inhibition, but may also increase the risk of cardiovascular events.

K, ketorol<u>ac</u>, is an injectable NSAID that helps migraines or severe pain. With a high risk of GI bleed, we limit its use to 5 days total of the oral and injectable dosage forms. For example, using two injection days and three days of the oral form makes five days total. **Ketorol<u>ac</u>'s Toradol** takes the middle part of the generic name and adds -dol to make it sound like 'dull' the pain.

IN, injectable, reminds us that ketorolac is an injectable.

From NSAIDs, we move to non-narcotic acetaminophen analgesics. Note that generic acetaminophen is generic paracetamol in many countries.

Question 19. Name three acetaminophen-containing analgesics or analgesic combinations and explain the role of each ingredient. To which organ does acetaminophen cause toxicity?

ABATE HEADACHE

A_____ (non-narcotic analgesic)

B_____ A_____ C_____

(barbiturate / non-narcotic analgesic / stimulant)

A_____ A_____ C_____

(NSAID / non-narcotic analgesic / stimulant)

Toxic to L_____ (organ)

E

19. ACETAMINOPHEN ANALGESICS - ABATE HEADACHE

> **A**cetaminophen [APAP] (Tylenol)
> **B**utalbital / Acetaminophen / Caffeine (Fioricet)
> **A**spirin / Acetaminophen / Caffeine (Excedrin Migraine)
> **T**oxic to liver
> **E**

QUICK SUMMARY

A, acetaminophen, provides analgesia, aspirin alleviates pain and inflammation, and caffeine vasoconstricts, speeding up metabolism and increasing onset of action.

Butalbital adds a stronger medicine for severe headaches. All three help ABATE a HEADACHE, our mnemonic.

Acetaminophen, is a non-opioid analgesic that reduces pain and fever. Unlike NSAIDs, APAP does not affect inflammation.

The generic **acetaminophen**, acronym **APAP**, and brand **Tylenol** all come from the chemical name:

N-<u>ace</u>tyl-para-<u>amino</u>-<u>phen</u>ol (**Acetaminophen**)
N-<u>A</u>cetyl-<u>P</u>ara-<u>A</u>mino-<u>P</u>henol (**APAP**)
N-ace<u>tyl</u>-para-amino-ph<u>enol</u> (**Tylenol**)

B, butalbital, as part of the combination butalbital / acetaminophen / caffeine also helps migraines. The prescription-only butalbital combination includes a barbiturate

81

that relaxes smooth muscle involved in headaches. Use the double "all" sound in **butalbital** to recognize this drug gets rid of "all" the headache pain.

A, aspirin, is part of the OTC Excedrin Migraine combination product, aspirin / acetaminophen / caffeine. We add the aspirin for inflammation and caffeine to vasoconstrict, quicken metabolism, and raise onset of action.

T, Toxic, is 'toxic to the liver,' because acetaminophen can be hepatotoxic in liver failure patients, alcoholics, or an overdose. To avoid hepatotoxicity, prescribers limit total daily dose to 4,000 mg. However, now recommendations include a 3,000 mg daily max.

Acetaminophen comes OTC alone or in combination, and we should look at total daily acetaminophen content in all products a patient takes to ensure we don't exceed the daily dose limits.

From over-the-counter and prescription acetaminophen analgesics to ABATE HEADACHE, we now move to opioids for severe pain.

Question 20. Name eight DEA Schedule II opioids or opioid combinations.

MY MY MY FREAKING HEAD HURTS OUCH OUCH

M_____ (opioid)

M_____ (opioid)

M_____ (opioid)

F_____ (opioid)

H_____ / A_____ (opioid / analgesic)

H_____ / I_____ (opioid / NSAID)

O_____ (opioid)

O_____ / A_____ (opioid / analgesic)

20. CII Opioids – MY MY MY FREAKING HEAD HURTS OUCH OUCH

> **M**orphine (Kadian, MS Contin)
>
> **M**eperidine (Demerol)
>
> **M**ethadone (Dolophine)
>
> **F**entanyl (Duragesic, Sublimaze)
>
>
> **H**ydrocodone / acetaminophen (Vicodin, Norco)
>
> **H**ydrocodone / Ibu<u>profen</u> (Vicoprofen)
>
> **O**xycodone (OxyIR, OxyContin)
>
> **O**xycodone / acetaminophen (Percocet)

QUICK SUMMARY

Opioids work on opioid receptors to reduce pain, stop coughing, and provide other therapeutic effects. Opiates and other controlled substances have DEA schedule numbers from Roman numerals one to five. Schedule one controls (C-I), like heroin, have no medically approved use. Opioids are schedule two (C-II) medications with strict laws.

They don't teach Roman numerals anymore in grade school, so these might be unfamiliar. To remember them, you can use the mnemonic "I Value Xylophones Like Cows Dig Milk," where I represents one, V, 5; X, 10; L, 50; C, 100; D, 500; and M, the number 1000.

Schedule II opioids have a high potential for physiological or physical dependence and abuse. Opioids effectively manage severe pain, especially in the hospital, but can be addictive and fatal at high doses. Opioids exert their analgesic effect by binding to opioid receptors, preventing pain signals from reaching the brain.

Opioids' danger usually comes from respiratory depression in overdose. Opioids cause constipation, so we'll pair them with a stool softener like docusate sodium or a stimulant laxative like senna. Other side effects include drowsiness and upset stomach. Some side-effects, like respiratory depression, exhibit tolerance, others do not.

The mnemonic MY MY MY FREAKING HEAD HURTS OUCH OUCH represents these eight compounds. The repeating words denote multiple drugs with the same first letter and the stuttering a patient in severe pain might have.

Here's a quick story about the difference between a computer test and real life. A student asked about the pain scale. An exam question read, "The patient reports pain at a ten on a scale of one to ten." The answers tested the speed at which a medication should run, either by mouth or intravenous.

I asked the student what the patient would probably say instead of "reporting pain at a ten." She said, "I think the patient would be screaming and need an injection or IV."

She was right. On exam questions, put yourself there instead of reading the scenario static on the page.

M, morphine, is a relatively long-acting opioid, available orally and by IV, that decreases myocardial oxygen demand through vasodilation. Reducing demand helps chest pain in acute coronary syndrome.

The generic name **morphine** comes from the ancient Greek god of dreams, Morpheus. The brand **Kadian** might come from cir**cadian** (the twenty-four-hour cycle) because **Kadian** is an extended-release morphine formulation. **M-S Contin** stands for **m**orphine **s**ulfate **contin**uous release, referencing the M-S from the beginnings of "morphine sulfate" and the "Contin" from the start of "continuous" – **M-S Contin**.

M, methadone, helps with opioid use disorder. Patients receiving opioid use disorder meds must visit treatment clinics daily. Methadone can be dangerous in opioid naïve patients. It has a long and unpredictable half-life, between 5 and 7 days, and patients may not know how they respond. Methadone can also cause QT prolongation. **Methadone**'s primary purpose is to help patients addicted to opiates get off narcotics, not provide pain relief.

M, meperidine, is short-acting and can lead to renal failure. Accumulation can lead to seizures. Meperidine is a drug we see most often in labor. Remember meperidine's brand **Demerol** as either "demolishing" or "dem-inishing" pain. Demerol.

F, fentanyl, is available in intravenous, patch, sublingual, and nasal dosage forms. When converting patients from other opioids to fentanyl, use the lowest fentanyl dose possible. Fentanyl is a synthetic and 100x more potent than heroin, so we have to be very careful with micrograms vs. milligrams.

When it was time to give our three-month-old preemie **fentanyl** for her pyloric valve stenosis surgery in the Pediatric Intensive

Care Unit (PICU) [[pick-you]], I made sure to check the calculated dose. **Duragesic** is a long **dura**tion anal**gesic** and comes in a patch that provides relief for 72 hours. Our Duragesic mnemonic takes the "dura" from long "duration" and "gesic" from "analgesic." **Sublimaze** is an injectable form of **fentanyl**.

H, hydrocodone / acetaminophen has its maximum hydrocodone daily dose limited by acetaminophen quantities.

H, hydrocodone / ibuprofen comes as a combination pain reliever plus anti-inflammatory. Brand **Vicoprofen** is like **Vicodin**, but with **ibuprofen** instead of **acetaminophen**. The manufacturer replaced "i-b-u" of **ibuprofen** with the "V-i-c-o" of **Vicodin**.

O, oxycodone comes as immediate-release and extended release formulations alone or combined with acetaminophen. The oxycodone molecule has one more oxygen atom than hydrocodone chemically.

OxyContin, the extended-release oxycodone-only form, has an abuse-deterrent formulation that makes it harder to crush and dissolve for illicit use.

O, oxycodone / acetaminophen. The "cet" in **Percocet** comes from a**cet**aminophen. Some students use that "codone" and "codeine" look a little alike to remember the similarity. However, Percocet contains oxycodone, not codeine. Drug companies add **acetaminophen** to Percocet as a mild analgesic.

Let's look at some specific morphine side effects next.

Memorizing Pharmacology Mnemonics

Question 21. Name the primary receptor we associate with opioid effects and seven possible opioid side-effects.

MORPHINE

M_____ (Receptor)

Possible opioid side-effects

O_____ S_____

R_____ D_____

P_____ P_____ (M_____)

H_____ , O_____

I_____ / C_____

N_____ / E_____

E_____

21. Morphine Side-Effects - **MORPHINE**

> ***M**u* receptor
>
> ***O**pioid sedation*
>
> ***R**espiratory depression*
>
> ***P**inpoint pupils (miosis)*
>
> ***H**ypotension, orthostatic*
>
> ***I**mpaction / constipation*
>
> ***N**ausea / emesis*
>
> ***E**uphoria*

QUICK SUMMARY

The MORPHINE mnemonic represents this opioid's characteristics and side effects. Patients can develop tolerance to some side-effects like respiratory depression. A non-tolerant individual might die from a certain dose, but a tolerant patient would notice little effect.

We move from DEA Schedule II morphine to CIII and CIV pain relievers in the next slide.

Question 22. Name one combination DEA Schedule III analgesic with acetaminophen and two DEA Schedule IV analgesics or combinations containing tramadol.

A/C TRAM WRECK

Acetaminophen w/ C_____ (Schedule III)

T_____ (Schedule IV)

T_____ with A_____ (Schedule IV)

22. CIII / CIV Opioids – A/C TRAM WRECK

> *DEA Schedule III*
> **A**cetaminophen /
> **C**odeine (Tylenol #3)
>
> *DEA Schedule IV*
> **T**ram<u>ad</u>ol (Ultram)
> **T**ram<u>ad</u>ol / acetaminophen (Ultracet)

Quick Summary

I thought of air-conditioned trams, A/C TRAMS, as providing relief, much as these drugs relieve pain. The trams' WRECK is why someone needs an analgesic. By placing a DEA Schedule III drug before DEA Schedule IV, that order makes it easier.

A, acetaminophen with codeine and tramadol products, provides analgesia. Schedule III opioids have less potential for abuse than schedule I or II substances, but abuse could lead to high psychological dependence or low physical dependence. Schedule IV opioids have a low potential for abuse relative to substances in schedule III.

C, codeine, is a prodrug the body must convert to morphine via CYP 2D6 in the liver for activation. A patient's ability to metabolize codeine varies. Some patients are ultra-rapid metabolizers who receive no benefit from the medication. Other patients are slow metabolizers whose slower metabolisms could lead to overdose. I am not sure why, but when you say the generic name of **Vicodin**, you say "**hydrocodone** *with* **acetaminophen**," but when you talk about **codeine,** you reverse the order as "**acetaminophen** *with* **codeine."** Students remember both because of the reverse order. The "#3" [[read # as "number"]] in **Tylenol #3** refers to the amount of codeine in combination. For example:

Tylenol #2 has 15 mg codeine with 300 mg acetaminophen.

Tylenol #3 has 30 mg codeine with 300 mg acetaminophen.

Tylenol #4 has 60 mg codeine with 300 mg acetaminophen.

T, tramadol, is a Schedule IV narcotic available by itself or with acetaminophen. Tramadol's risks include seizure and serotonin syndrome, and you must dose it renally. Tramadol only weakly affects opioid receptors. Therefore, the DEA did not classify tramadol as a controlled substance until 2014. The "adol, a-d-o-l" stem indicates that it's a mixed opioid analgesic. Many students think of "tram wreck" and "train wreck" as a way to remember that **Ul*tram*** is for pain.

Tramadol with **acetaminophen** follows tramadol by itself. The manufacturer took the brand name **Ultram** (for tramadol alone), dropped the "m," and added "acet" from **acetaminophen** to make **Ultracet**. Usually, I would continue to DEA Schedule V after II, III, and IV, but you'll find those cough suppressants in the respiratory chapter.

From these opioid agonists, we move to the opposing opioid antagonists.

Question 23. Name two opioid antagonist medications and the receptor they affect.

OMNIBUS AGAINST OPIOIDS

O_____ A_____ (drug class)

M_____ R_____ (affected receptor)

N_____ (opioid antagonist)

I

B_____/N_____ (opioid partial agonist / opioid antagonist)

U

S

23. Opioid Antagonists – OMNIBUS AGAINST OPIOIDS

> **O**pioid antagonists
> **Mu** receptors
> **N**aloxone (Narcan)
> **I**
> **B**uprenorphine / naloxone (Suboxone)
> **U**
> **S**

Quick Summary

We use naloxone alone for acute opioid overdose while the buprenorphine / naloxone combination as Suboxone is for opioid withdrawal treatment and prevention. Note the "–nal, n-a-l" naloxone stem. When someone says omnibus, I remember the Omnibus Reconciliation Act, or COBRA. COBRA regulates health insurance after people move jobs. With the sentiment against opioids, I thought OMNIBUS AGAINST OPIOIDS made sense for opioid antagonists.

O, opioid antagonists, bind opioid receptors but don't activate them, blocking opioid responses. Opioids, while useful for pain, can have serious negative consequences.

M, *mu* *[pronounced myou]*, receptors are the primary opioid receptor.

N, na̲loxone, is an opioid antagonist that blocks opioid receptors without providing analgesia, constipation, or upset stomach. We use it to reverse respiratory depression from an opioid overdose. Paramedics and police officers often have this medication and many pharmacies can dispense it without a prescription. It has IV, SubQ injection, and intranasal forms. With a short duration of action, patients who use naloxone should immediately seek care to avoid repeat withdrawal. Buprenorphine with naloxone protects against abuse. The **naloxone** keeps patients from crushing and injecting the drug.

The "nal, n-a-l" stem in **naloxone** indicates it's an opioid receptor antagonist, but the brand name **Narcan** with its elements "Narc" and "an," also hints at a na̲rcotic a̲ntagonist.

Naloxone is an opioid receptor antagonist used in opioid overdose situations. Often, it's in the L-E-A-N, lean, acronym of emergency medicines **l̲idocaine**, **e̲pinephrine**, **a̲tropine**, and **n̲aloxone**.

B, buprenorphine / na̲loxone. Buprenorphine is a partial opioid agonist, which binds opioid receptors so opioids cannot. Some use buprenorphine for pain and others for opioid addiction. It binds opioid receptors, and does not exert an effect or cause withdrawal. Patients use **buprenorphine / naloxone** like **methadone** to help detox from opiate addiction. The **naloxone** keeps patients from crushing and injecting it.

Now, we turn to opioid antagonists used for opioid-induced constipation.

Question 24. Name three medications for opioid-induced constipation.

THREE MEDICINALS

M_____

N_____

N_____

24. Opioid Antagonists for Opioid-Induced Constipation (OIC) – THREE MEDICINALS

> **M**ethyl**nal**trexone (Relistor)
> **N**al**demedine (Symproic)
> **N**al**oxegol (Movantik)

Quick Summary

These drugs work to prevent opioid induced constipation, (OIC) and each have the n-a-l, nal stem at the end of THREE MEDICINALS.

M, methylnaltrexone, is a subcutaneous injection and opioid antagonist. It's a naltrexone derivative that, with the extra methyl group, can't cross the blood-brain barrier and reduce opioid analgesic effects. The brand name **Relistor** combines "relieve" and "restore" as in relieve constipation and restore natural bowel movements.

N, naldemedine, is another opioid antagonist for OIC, for patients with chronic non-cancer pain. In comparison to regular laxatives, patients take this once daily irrespective of food.

N, naloxegol, is also for OIC. Patients use once daily on an empty stomach. Withdrawal symptoms and severe diarrhea are possible. Naloxegol's brand name includes the first three letters of "move" as in a bowel "movement." **Movantik**.

Now we take a look at 5-HT$_1$ agonists in the next slide.

Memorizing Pharmacology Mnemonics

Question 25. Name seven triptans for a migraine and at least three dosage forms.

FEARS ZIN HEADACHES

F_____ (5HT1 agonist)

E_____ (5HT1 agonist)

A_____ (5HT1 agonist)

R_____ (5HT1 agonist)

S_____ (5HT1 agonist)

Z_____ (5HT1 agonist)

I_____, N_____ S_____,
O_____ D_____T_____
(three dosage forms)

N_____ (5HT1 agonist)

98

25. 5-HT₁ Agonists – FEARS ZIN HEADACHES

> **F**rova<u>triptan</u> (Frova)
> **E**le<u>triptan</u> (Relpax)
> **A**lmo<u>triptan</u> (Axert)
> **R**iza<u>triptan</u> (Maxalt)
> **S**uma<u>triptan</u> (Imitrex)
>
> **Z**olmi<u>triptan</u> (Zomig)
> **I**njections, nasal sprays, and orally disintegrating tabs (ODT) available for some triptans
> **N**ara<u>triptan</u> (Amerge)

Quick Summary

Migraines, unlike wine headaches, are debilitating. 5-HT₁ agonists, triptans, are first-line migraine onset treatments. Watch for the "–triptan, t-r-i-p-t-a-n" stem. Students remember the class suffix "–triptan" "trips up" a headache. I often confused triptans as agonists or antagonists. Use migraine "agony" to remember agonist. We dose triptans once or twice at migraine onset, as excessive use can result in rebound headaches. Most triptans have multiple dosage form routes: subcutaneous injections, nasal sprays, and orally disintegrating tablets for patients who are nauseous and vomiting. If one triptan doesn't work, we can switch to another.

The mnemonic FEARS ZIN HEADACHES combines the "Zin" abbreviation for Zinfandel wine and headache together.

F, frovatriptan I'll discuss in detail with naratriptan below. Use the "ova, o-v-a" from **frovatriptan** or **Frova** as in the migraine is "ova'" or "over."

E, eletriptan, undergoes CYP 3A4 metabolism in the liver and we monitor for drug-drug interactions. Eletriptan's brand name **Relpax** combines "Rel" for "relief" and "pax," the Latin word for "peace."

A, Almotriptan brand **Axert** makes you think "that ax hurt" for migraine.

R, rizatriptan has an orally disintegrating tablet form. Rizatriptan will help you "rise" from the pain.

S, sumatriptan, is the only first-generation triptan. With a shorter half-life and oral bioavailability, this could have implications for headache recurrence. Sumatriptan has a STATdose Pen form for patients that can't get down oral meds.

Z, zolmitriptan.

I, injections, nasal sprays, and orally disintegrating tabs (ODT) available for some triptans. This is necessary for nauseous and vomiting patients.

N, naratriptan and **frovatriptan** have slow onsets of action and are long-acting. They're a good option for long flights or car rides, resulting in fewer recurring headaches. **Naratriptan's Amerge** will let you "emerge" from that dark quiet room.

Moving down from the head to joints, we tackle the DMARDs.

Question 26. Name four non-biologic DMARDs.

MASH FOR JOINTS

M_____

A_____

S_____

H_____

26. Disease Modifying Anti-Rheumatic Drugs (DMARDS) – Non-Biologics – MASH FOR JOINTS

> **M**etho<u>trexate</u> (Rheumatrex)
>
> **A**zathioprine (Imuran)
>
> **S**ulfa<u>sal</u>azine (Azulfidine)
>
> **H**ydroxychloroquine (Plaquenil)

Quick Summary

Prescribers often employ disease-modifying antirheumatic drugs (DMARDs) against rheumatoid arthritis, RA. Rheumatoid arthritis is an inflammatory disease with weakness, malaise, and joint involvement, especially joint stiffness upon waking.

These DMARD non-biologics came first, although biologics are often first-line therapy. The mnemonic MASH FOR JOINTS starts with the Mobile Army Surgery Hospital acronym "MASH, M-A-S-H."

M, metho<u>trexate</u>, is the gold-standard DMARD for RA. It can cause folic acid deficiency requiring supplements. Patients can experience nausea, vomiting, stomatitis, and liver enzyme elevations. This medicine is toxic to the fetus. Note, when a patient can't tolerate methotrexate, we might see leflunomide.

Methotrexate's "trexate, t-r-e-x-a-t-e" stem helps remind you this is a DMARD. One student came up with "Meth o T-Rex ate the rheumatic inflammate" – Methotrexate. The "rheuma" in the brand name **Rheumatrex** reminds you it relieves rheumatoid arthritis.

A, azathioprine, like methotrexate, has undesirable GI side effects. Azathioprine causes leukopenia and doesn't work as well as methotrexate. Azathioprine's **Imuran** refers to the immune system with "imu, i-m-u" and rheumatoid arthritis with "ra, r-a."

S, sulfasalazine, is a prodrug cleaved into two active products. Unlike methotrexate, sulfasalazine is safer with pregnancy. Still, nausea, vomiting, and rash limit its use. **Sulfasalazine** has two stems, "sal, s-a-l" for salicylate and "sulfa, s-u-l-f-a" for a sulfa moiety.

H, hydroxychloroquine, takes 3-6 months to activate and is one of the least productive DMARDs. Overall, it has a mild side effects, but rare retinal damage necessitate annual eye exams.

Next, we move to the DMARD biologics.

Question 27. Name five biologic DMARDs.

GO ADD IN A BETTER DMARD

Go_____

Ad_____

In_____

Ab_____

Et_____

27. DMARDs - Biologics – GO ADD IN A BETTER DMARD

> *GO*limu<u>mab</u> (Simponi)
> *AD*alimu<u>mab</u> (Humira)
> *IN*fli<u>ximab</u> (Remicade)
> *AB*a<u>tacept</u> (Orencia)
> *ET*a<u>nercept</u> (Enbrel)

Quick Summary

Sometimes a cheaper non-biologic fails to relieve rheumatoid arthritis pain and symptoms, and we need to GO ADD IN A BETTER DMARD. By taking the first two letters of each biologic DMARD, the mnemonic readily brings complex drug names to mind. Note how the biologics' stems form. The "–mab, m-a-b" means monoclonal antibody, but the other parts, the sub-stems, have meaning, as well. For example, the "li, l-i," in infliximab means immunomodulator. The "xi, x-i," refers to its chimeric origin.

Biologics often affect the immune system, increasing infection risk. These newer genetically engineered drugs often come at a higher price point but typically work faster, at 6-8 weeks, than non-biologics. Crohn's disease and lupus also respond to them.

GO, golimu<u>mab</u>, is a TNF-alpha inhibitor neutralizing TNF mediated inflammation with a subcutaneous once monthly dosage. Short-term side effects include injection site reaction and infection risk, but we still need long-term data.

AD, adalimumab, is a TNF-alpha inhibitor with every other week dosing. It's either a first-line RA medicine or for patients with severe RA with poor response to other DMARDs. Adalimumab may cause confusion, infection, injection site reactions, and paresthesia.

IN, infliximab, is an IV TNF-alpha inhibitor only approved for combination therapy with methotrexate. Headaches come frequently and developing infliximab antibodies can cause a lupus-like reaction. After initial titration, infusions come every eight weeks.

AB, abatacept, prevents stimulation of T-cell activation and may cause pulmonary side effects, so be careful with asthma and COPD patients. Abatacept has a complex stem. The "-ta-" infix in **abatacept** means it's going after T-cell receptors and the suffix "–cept" means that it's a re**cept**or molecule, either native or modified.

ET, etanercept, is a TNF-alpha inhibitor, given once or twice weekly. Avoid in patients with a serious infection risk. The "-ner-" infix in **etanercept** points out that it goes after tumor necrosis factor receptors. The "–cept," as with abatacept, means that it's a re**cept**or molecule, either native or modified

Side note:

Sarilumab (Kevzara) is another RA medication for patients that failed previous DMARDs.

From joints, we move to bones.

Question 28. Name four bisphosphonates, a specific fracture concern, and instructions for the patient's positioning.

FRAZIL BONES

F_____ (fracture concern)

R_____ (bisphosphonate)

A_____ (bisphosphonate)

Z_____ (bisphosphonate)

I_____ (bisphosphonate)

L_____ (patient's positioning prohibition)

Memorizing Pharmacology Mnemonics

28. BISPHOSPHONATES – FRAZIL BONES

> **F**ractures, atypical of the femur
>
> **R**ise<u>dronate</u> (Actonel)
>
> **A**len<u>dronate</u> (Fosamax)
>
> **Z**oledronic acid (Reclast, Zometa)
>
> **I**ban<u>dronate</u> (Boniva)
>
> **L**ying down prohibited for 30 to 60 minutes

QUICK SUMMARY

Bisphosphonates prevent bone density loss in osteoporosis patients. Watch for dyspepsia, a common side effect of all bisphosphonates. We can mitigate dyspepsia risk or reflux with proper administration. A rare side effect, osteonecrosis of the jaw, happens with patients receiving very high dose IV bisphosphonate therapy, especially in cancer patients.

Memorize these as calcium metabolism regulators from the "dronate, d-r-o-n-a-t-e" stem. Students like to remember that "drone" rhymes with bone.

When you have "frazil, f-r-a-z-i-l" ice, it's slushy because the unsettled water prevents freezing. Fragile osteoporosis bones resemble the mnemonic FRAZIL BONES.

F, femur, references the rare atypical femur fracture.

R, rise<u>dronate</u>, we can dose daily, weekly, or monthly. Risedronate has a delayed release mechanism, so patients should take it after breakfast. Risedronate's **Actonel** "acts on" bone.

A, alendronate, comes in a once daily or once weekly form. Many students have said that alendronate's brand name **Fosamax** looks like a "fossil."

Z, zoledronic acid, is an annual IV infusion and commonly treats hypercalcemia often caused by malignancy. **Zoledronic acid** is also known as **zoledronate**, which has the bisphosphonate stem. Its brand name, **Reclast**, refers to "reverse" and "osteoclast" as its mechanism of action is to reverse or block osteoclasts.

I, ibandronate, has an oral monthly dosage form or every three months IV form. Ibandronate's **Boniva** has the first three letters of "bone" and middle two letters of "live" as in "bones live." The generic **ibandronate** contains all the letters in **Boniva** except the "v."

L, lying down prohibited for 30 to 60 minutes. The dyspepsia risk warrants that patients take bisphosphonates first thing in the morning before breakfast with a full glass of water. Patients should not lay down; instead, they should remain upright afterward, either sitting or standing, for 30 to 60 minutes, depending on the bisphosphonate.

Side notes:

Pamidronate IV, and **Zoledronic acid** are for bone pain and to prevent bone loss from bone metastasis cancer. We can use both in non-cancer patients, but they are preferred agents in cancer patients.

There's a monoclonal antibody, **denosumab**, that's a RANKL inhibitor, administered subQ every six months for osteoporosis and subQ every four weeks in patients with bone metastasis. This therapy is for patients resistant or intolerant to bisphosphonates.

Abaloparatide (Tymlos), a parathyroid hormone analog for osteoporosis, slows bone degradation and increases bone density. Approval only exists for postmenopausal women with a high bone fracture risk. **Teriparatide (Forteo)** is also a parathyroid hormone analog that men and postmenopausal women with osteoporosis can use.

From bone, we move to muscle relaxation.

Question 29. Name seven muscle relaxants.

CAR MET CYCLE = BACK TO MD (MEDIC)

Car_____

Met_____

Cyc_____

Bac_____

T_____

Me_____

Di_____ (benzodiazepine)

29. Muscle Relaxants – CAR MET CYCLE = BACK TO MD

> **C**arisoprodol (Soma)
>
> **Met**axalone (Skelaxin)
>
> **Cycl**obenzaprine (Flexeril)
>
> **B**aclofen (Lioresal)
>
> **T**izanidine (Zanaflex)
>
> **M**ethocarbamol (Robaxin)
>
> **D**iazepam (Valium)

QUICK SUMMARY

Muscle relaxers reduce pain and stiffness but do not provide anti-inflammatory action. A story mnemonic like this one represents a win to me. When the CAR MET the biCYCLE, that's when the person had to go BACK TO the MD.

Muscle relaxants, or antispasmodics, are our next category of medications, and they work to relax or reduce muscle tension.

CAR, carisoprodol, is a centrally acting schedule IV muscle relaxer with a risk of withdrawal, tolerance, and dependence. It causes drowsiness, so along with preventing spasms, it can help patients asleep. CYP 2C19 metabolizes carisoprodol in the liver, and, like codeine, affects patients differently depending on whether the patient is a rapid or slow metabolizer. We often see carisoprodol abused especially with alcohol. Some of the brand names hint at muscle relaxation – for example, **carisoprodol**'s brand **Soma** sounds like so<u>m</u>nolence, which means sleepiness. If you are a literature nerd, Soma was a hallucinogen in Aldous Huxley's 1932 book, *Brave New World*.

MET, metaxalone, exerts its muscle relaxant effect through sedation. It can be hepatotoxic. **Skelaxin**, the brand name of metaxalone, alludes to "<u>skel</u>etal rel<u>axin</u>'."

CYCL, cyclobenzaprine, is both a muscle relaxant and anticholinergic, so watch for dry mouth and sedation. Caution patients taking SSRIs, SNRIs, or TCAs due to the serotonin syndrome risk. Because of this serotonin syndrome risk, avoid combining cyclobenzaprine with MAOIs. Cyclobenzaprine helps you get bending again, using the 'b-e-n-z' in the center of the generic name. The brand *Flexeril* improves *flex*ibility.

BAC, baclofen, comes in oral and intrathecal forms. Abrupt intrathecal baclofen withdrawal can lead to rebound spasticity and rhabdomyolysis, which can lead to organ failure. With its sedative properties, we use it cautiously in the elderly. With **baclofen**, turn around the "l-o, lo" and "b-a-c-, bac" from baclofen to remember "low back" pain.

T, tizanidine, is a centrally acting alpha-2 agonist that can lead to hypotension. Taper if discontinuing to avoid rebound hypertension and tachycardia. Tizanidine has anticholinergic properties, so it causes dry mouth. **Zanaflex,** tizanidine's brand name, ends with "flex" for increased fle<u>x</u>ibility.

O.

Me, methocarbamol, is oral or injectable. With less sedation, it's suitable for daytime use. However, it can cause urine discoloration and hypotension. **Robaxin,** the brand name of methocarbamol**,** and rel<u>axin</u>' go together.

Di, di<u>azepam</u>, a benzodiazepine, comes orally or by injection, either intravenously or intramuscularly. It is DEA Schedule IV and causes sedation. For seizures, we see a rectal form because a seizing patient can't take medicine by mouth.

C.

Diazepam and benzo**diazep**ine, diazepam's drug class, have similar letters. Note the "v-a-l" in **Val**erian root, which is an herbal remedy for anxiety and has the same three initial letters as **Valium**. Some think of **Valium** as relaxing anxiety and muscles similarly. The "–azepam, a-z-e-p-a-m" stem in diazepam identifies it as a benzodiazepine.

From muscle, I move down to the toes, where half of gouty arthritic inflammations happen.

Question 30. Name a uricosuric, two xanthine oxidase inhibitors, an NSAID for gout, a second line agent for gout pain, and a gout medicine for refractory gout treatment.

PACIFY PEG'S TOE

P_____ (uricosuric)

A_____ (xanthine oxidase inhibitor)

C_____ (second-line agent for gout pain)

I_____ (NSAID)

F_____ (xanthine oxidase inhibitor)

Y

PEG_____ (refractory gout treatment)

30. ANTI-GOUT AGENTS – PACIFY PEG'S TOE

> **P**robenecid (Benemid)
> **A**llopurinol (Zyloprim)
> **C**olchicine (Colcrys)
> **I**ndo<u>methacin</u> (Indocin)
> **F**ebu<u>xostat</u> (Uloric)
> **Y**
>
> **P**<u>eg</u>loticase (Krystexxa)

QUICK SUMMARY

Gout is the oversecretion or underexcretion of uric acid, which leads to uric acid crystals accumulation in tissues and around joints. Often, painful gout attacks start in the big toe and spread upwards to other joints. Our mnemonic is PACIFY PEG's TOE.

P, probenecid, is a uricosuric agent that increases renal clearance of uric acid contraindicated with renal dysfunction. Side effects include rash and GI upset. Probenecid can precipitate an acute gout attack. As such, patients should remain hydrated, as uric acid stones may form from its mechanism of action. **Probenecid** ends in ecid, e-c-i-d which is much like uric acid, a-c-i-d. It's a "pro" that "ben-efits" the uric "acid" patient.

A, allopurinol, a xanthine oxidase inhibitor we'll address with febuxostat below. Within allopurinol, you can see "uri, u-r-i," which corresponds to the uric acid the medication reduces. Remember this is an anti-arthritic by thinking of joints becoming "all-pure-and-all," allopurinol.

C, colchicine relieves acute gout attacks. The prescriber provides one tablet at onset followed by one tablet one hour later if the patient can tolerate the unpleasant dose-related GI side effects. Colchicine works best within the first 24 hours of acute attack onset. Colchicine has many drug interactions, and we renally dose it. Because of all these complications, colchicine is a 2nd line behind NSAIDs for acute attack and needs renal adjustment. Colchicine relieves an acute gouty attack, so I put that before the uric acid reducers that treat chronic increased uric acid. The "crys" in **Colcrys** sounds like the painful gouty crystals that often form in the big toe. Google "gouty crystal" images, and you'll see that they look like needles.

I, Indomethacin, is a long-acting drug we see most often when choosing an NSAID as treatment.

F, febuxostat, like allopurinol, is a xanthine oxidase inhibitor that works by inhibiting uric acid production. Both are first-line acute gout attack prevention treatments and help uric acid underexcreters or overproducers. Don't stop this medicine during an acute attack. Allopurinol is relatively inexpensive, while febuxostat, newer to the market, and costs more. Allopurinol can cause a rash, and both medications can produce GI upset and elevated liver function tests (LFTs). Avoid febuxostat in heart disease and heart failure patients. We can also see rebound hyperuricemia.

The "xostat, x-o-s-t-a-t" stem in febuxostat indicates a xanthine oxidase inhibitor that prevents uric acid from forming. The "x" and "o" in the stem included in the generic name match these first letters in xanthine oxidase. The brand name **Uloric** looks like "U" "lower" "uric acid" – Uloric.

P, pegloticase, reduces serum uric acid levels and helps with gout prevention for patients with severe disease refractory to other treatments. We premedicate patients with diphenhydramine, acetaminophen, hydrocortisone, and hydration before infusion to avoid reactions. It can cause rebound hyperuricemia and is very expensive. Pegloticase's brand **Krystexxa** has the first five letters of "crystals" if you substitute the "k" with a "c" and "ex, e-x" as exit. Uric crystals exit.

Side note:

Rasburicase, similar to pegloticase in that it converts uric acid to allantoin, is only different in that pegloticase is PEGylated, increasing the half-life and decreasing immune reaction chance.

From the musculoskeletal medications, we move to the respiratory chapter.

CHAPTER 3: RESPIRATORY MNEMONIC FLASHCARDS

Memorizing Pharmacology Mnemonics

Question 31. Name five first-generation antihistamines and two side-effects of concern.

ALLERGY ABCD RHYME

A_____ (side effect mechanism)

B_____ (antihistamine)

C_____ (antihistamine)

D_____ (antihistamine)

R_____, S_____ (side effect concern)

Hy_____ (antihistamine for hives)

Me_____ (antihistamine for vertigo)

31. 1ST GENERATION ANTIHISTAMINES – ALLERGY ABCD RHYME

> **A**nticholinergic
>
> **B**rompheniramine (Bromfed, Dimetapp)
>
> **C**hlorpheniramine (Chlorphen)
>
> **D**iphenhydramine (Benadryl)
>
> **R**est, sedation causing
>
> **H**ydroxyzine (Vistaril)
>
> **M**eclizine (Antivert)

QUICK SUMMARY

Oral antihistamines are first-line drugs for allergic rhinitis and can reduce symptoms of sneezing and itching and runny nose. Antihistamines don't affect congestion. They selectively bind histamine-1 receptors to prevent these symptoms. There are two types of antihistamines – first-generation, or non-selective antihistamines, and second-generation, or selective antihistamines.

First-generation antihistamines are 'older' antihistamines more likely to cause drowsiness and cognitive side effects, especially in the elderly. Sedation occurs because first generation antihistamines can cross the blood-brain barrier and bind to the brain's histamine receptors.

The ALLERGY ABCD RHYME mnemonic includes the first letters of these antihistamines.

A, anticholinergic, reminds us of the ABDUCT WATER mnemonic including urinary retention, constipation, and dry mouth.

B, brompheniramine, and

C, chlorpheniramine, are both available in over-the-counter (OTC) tablets, liquids, and syrups. They have a little sedative effect, but are moderately anticholinergic and can cause dry mouth, blurred vision, and constipation. Brompheniramine and chlorpheniramine both replace the "ine, -i-n-e" from their elemental forms on the Periodic Table of Elements bromine and chlorine with -pheniramine. If you remove the p, e, and n, from –pheniramine and replace the "r" with s-t, you form the word histamine.

D, diphenhydramine, also OTC, uses its sedative effect therapeutically helping patient's sleep. It's in cough and cold products and combined with acetaminophen in Tylenol PM. Patients reach for diphenhydramine during allergy season or for allergic skin reactions. Diphenhydramine, like brompheniramine and chlorpheniramine, has anticholinergic effects. Remember diphenhydramine's brand name by recognizing **Ben**adryl <u>ben</u>efits you by <u>dry</u>ing up your runny nose, taking the "ben" from "benefits" and the "dry" from "drying." Some students associate **Benadryl's** capital "B" with the blood-brain barrier (BBB), which **Benadryl** *can* pass through.

R, rest, reminds us of sedation. However, hangover effects are common.

HY, hydroxyzine, is a prescription-only antihistamine for itching secondary to hives and dermatitis. We give hydroxyzine orally for general anxiety disorder or intramuscularly, IM, for hospitalized anxiety patients. Like first-generation antihistamines, hydroxyzine has drowsiness and anticholinergic side effects. Watch for look-alike sound-alike (LASA) with hydralazine, a vasodilator.

ME, meclizine's primary role is in preventing nausea and vomiting associated with motion sickness and vertigo. Take it 60 minutes before travel. Meclizine has anticholinergic side effects but is less sedating than diphenhydramine.

Side notes:

Two other related drugs include **orphenadrine** and **azelastine**. Orphenadrine is for skeletal muscle relaxation, but structurally related to diphenhydramine. **Azelastine**, brand **Astelin**, is an antihistamine nasal spray.

After the first generation, look to the less-sedating second generation.

Memorizing Pharmacology Mnemonics

Question 32. Name four second-generation antihistamines and the drug class' primary advantage over the first-generation. Provide one class suffix.

CALLED FOR NON-DROWSY

C_____ (antihistamine)

A_____ (primary advantage)

L_____ (left enantiomer antihistamine)

L_____ (antihistamine)

E_____ (another advantage)

D_____ (active metabolite antihistamine)

F_____ (active metabolite antihistamine)

CLASS SUFFIX -a_____

32. 2ND GENERATION ANTIHISTAMINES – CALLED FOR NON-DROWSY

> **C**etirizine (Zyrtec)
> **A**void drowsiness
> **L**evocetirizine (Xyzal)
> **L**oratadine (Claritin)
> **E**ntering brain prohibited
> **D**esloratadine (Clarinex)
>
> **F**exofenadine (Allegra)

QUICK SUMMARY

We prefer second-generation antihistamines, or selective antihistamines, for allergic rhinitis because they're not as sedating as first-generation drugs. This reduced sedation comes from the drug's inability to cross the blood-brain barrier and cause fewer anticholinergic side effects. Our mnemonic is CALLED FOR NON-DROWSY.

C, cetirizine. I pronounce the "t-i-r" in cetirizine as a tear, like a teardrop. I think cetirizine protects me from tearing from allergies.

A, avoid drowsiness, the advantage of the second generation.

L, levocetirizine, the active left enantiomer of cetirizine, requires half the cetirizine dose. The commercials for the Xyzal owl mascot resonates, as my kids have three stuffed Xyzal owls that talk about the drug.

L, loratadine, has the '-atadine' stem, which marks it is an H_1 blocker. '-atadine' sounds similar to the '-tidine' stem in H_2 blockers for acid reflux. Don't confuse the two. The **Claritin** "clear" commercials resonate with *clearing* one's head from allergies or *clear* eyes relieved from allergies. Careful with **ClariSpray**, it is a brand extension that contains fluticasone, a steroid, not the antihistamine loratadine.

E, entering brain prohibited reminds us these drugs can't bypass the blood-brain barrier.

D, desloratadine, also has the '-atadine' stem, and is the active metabolite of loratadine, so a patient needs only half the loratadine dose. For example, 10 mg of loratadine is the same as 5 mg of desloratadine. **Clarinex** is the *next* update from **Claritin**.

F, fexofenadine, is another second-generation antihistamine, one we find in children's allergy products. The third-generation antihistamine fexofenadine is a safe, active metabolite of **terfenadine**. Fexofenadine's brand name **Allegra** takes six letters from the word "allergy" and changes them to an Italian word meaning "lively." Allegra. [[Italian accent]]

Side notes:

Why did I order cetirizine first, then levocetirizine and loratadine then desloratadine? Some consider levocetirizine a *third-generation* antihistamine and cetirizine a second-generation. Second precedes third numerically. Loratadine should precede desloratadine, as original molecules precede their metabolite. Fexofenadine, a metabolite, would have come after its parent compound, terfenadine. However, the manufacturer withdrew terfenadine for cardiac effects, so I've excluded it.

From antihistamines, we move to decongestants and stay in the nose.

Memorizing Pharmacology Mnemonics

Question 33. Name three decongestants including a decongestant/antihistamine combination. Outline the limit on days' use for intranasal decongestants and a disease state to avoid with these medicines.

STUFFED UP PEOPLE

P_____ (decongestant)

E_____ (days' use limit)

O_____ (decongestant)

P_____/ (decongestant with)

L_____ (antihistamine)

E_____ U_____ H_____ (disease state to avoid)

128

33. Decongestants – STUFFED UP PEOPLE

> **P**henylephrine (NeoSynephrine)
>
> **E**xceed three days prohibited (rebound congestion)
>
> **O**xymetazoline (Afrin)
>
> **P**seudoephe<u>drine</u> / (alone is Sudafed)
>
> **L**or<u>ata</u>dine (and pseudoephe<u>drine</u> is Claritin-D)
>
> **E**xclude uncontrolled hypertensive patients

Quick Summary

Decongestants are alpha agonists, or sympathomimetics that stimulate the sympathetic nervous system. Stimulating the SNS causes vasoconstriction of dilated sinus vessels reducing congestion.

Unfortunately, vasoconstriction may go further than the sinuses. Decongestants can affect the cardiovascular system as well. This raises blood pressure with tachycardia, and palpitations. These palpitations may lead to tremors and restlessness. Because decongestants can raise blood pressure, we avoid them in hypertensive patients. Locally-acting nasal forms are sometimes okay in cardiovascular disease.

Watch for hidden acetaminophen and overdose in multi-symptom products. The antihistamines in combination products are either second-generation for daytime or first-generation for nighttime.

The mnemonic STUFFED UP PEOPLE resembles 'stuck up people' with noses in the air.

P, phenylephrine, the weak cousin of pseudoephedrine, comes in intranasal and oral dosage forms. It doesn't have limits like pseudoephedrine and is available over-the-counter. It's less effective because of low bioavailability. Phenylephrine sounds like **pseudoephedrine** – they are both decongestants. Patients recognize pseudoephedrine by the "hyphen D" and phenylephrine by the "P-E" abbreviation in many cold preparations.

E, exceed three days prohibited, rebound congestion possible reminds us of safe limits of use.

O, oxymetazoline is an intranasal decongestant with a three day maximum to avoid rebound congestion. This spray directly constricts dilated mucosa, resulting in fewer systemic side effects. Local issues may include stinging, burning, and sneezing. The "**Afrin**" brand name sounds a like the "ephrine" from **phenyl*ephrine***, so you can relate these decongestants.

P, pseudoephedrine, alone is the oral decongestant brand Sudafed. It is a behind-the-counter medication that doesn't need a prescription, but patients must show identification. It's possible to make illicit methamphetamine from pseudoephedrine, so there are purchase limits per transaction, day, and month. Pseudoephedrine has dose-related blood pressure effects. If you take out the second "e" and drop the "rine" (r-i-n-e) from **pseudoephedrine**, you get the **Sudafed** pronunciation. One student of mine said she was p-h-e-d up, "phed up," with congestion and that's how she remembered it.

L, loratadine. When we add this antihistamine, the brand name becomes Claritin-D, where the D represents pseudoephedrine with its therapeutic effect, *D,* for decongestion. Therefore, adding **loratadine**, an antihistamine, to **pseudoephedrine**, a decongestant, helps with both runny and stuffy noses.

In the next slide, we reduce inflammation with intranasal steroids.

Question 34. Name four intranasal corticosteroids and a brand name often mistaken for an antihistamine.

BFF MOM KNOWS NOSE

B_____

F_____

F_____

Mom_____

Loratadine is brand Claritin but fluticasone is brand _____ which might be mistaken for an antihistamine

34. Intranasal Corticosteroids – BFF MOM KNOWS NOSE

> **B**udesonide (Rhinocort)
>
> **F**luticasone propionate (Flonase, Clarispray)
>
> **F**luticasone furoate (Veramyst)
>
> **M**ometasone (Nasonex)

Quick Summary

Intranasal corticosteroids work against seasonal or chronic allergy symptoms by helping with sneezing, runny nose, itching, and congestion. Most patients tolerate nasal corticosteroids well, but they can cause a headache, nasal irritation, or nosebleeds. The controversy surrounds long-term side effects commonly seen with oral steroids like poor wound healing, adrenal suppression, and infection. Intranasal steroids have little systemic absorption and fewer side effects.

These nasal corticosteroids take a few weeks to begin working, so patients start them *before* the allergy season. Failing to treat congestion might result in the condition worsening traveling down your throat developing into a cough.

The various intranasal corticosteroids have similar efficacy and make up the BFF MOM KNOWS NOSE mnemonic.

B, budesonide has no recognized stem, nor do **fluticasone**, or **mometasone**, but often "o-n" or "o-n-e," pronounced like I "own" something, matches the "o-n-e" at the end of

testosterone, a more familiar steroid. Budesonide's **Rhinocort** combines "rhino" meaning "nose" and "cort" which is short for corticosteroid.

F, fluticasone propionate's Flonase combines parts of "airflow" and "nasal" for the intranasal form. The Flonase name contrasts **Flovent**, which combines "airflow" and "ventilation" for the orally inhaled form.

F, fluticasone furoate, where the propionate and furoate indicate different ester sub-units in fluticasone's steroid backbone. **Fluticasone furoate's** brand **Veramyst** shortens "very fine mist."

MOM, mometasone's Nasonex combines nasal and ex, meaning "out" as in "congestion get out."

Side note:

Triamcinolone spray is another common intranasal corticosteroid I couldn't just work into the mnemonic. Triamcinolone's brand **Nasacort** combines "nasal" and "corticosteroid."

The next slide moves us from the nose into coughing.

RESPIRATORY MNEMONIC FLASHCARDS

Question 35. Name four anti-cough medications or combination products.

BIG DARN COUGH

B_____ (antitussive)

I

G_____ / (expectorant)

D_____ (antitussive)

CO_____ / (opioid)

U

G_____ (expectorant)

H_____ / C_____

(opioid / antihistamine)

135

35. Cough – BIG DARN COUGH

> **B**enzonatate (Tessalon Perles)
> **I**
> **G**uaifenesin / **D**extromethorphan (Mucinex DM, Robitussin DM)
> **CO**deine /
> **U**
> **G**uaifenesin (Cheratussin AC)
> **H**ydrocodone / Chlorpheniramine polistirex (Tussionex)

Quick Summary

MuciNEX, TUSSioNEX, and CheraTUSSin are three popular combination cough products. Each has a regulated medication: dextromethorphan, hydrocodone, or codeine.

B, benzonatate, works as a local anesthetic to decrease lung receptor sensitivity and cough. The "Tess" in **Tessalon Perles** looks like "tussive." Imagine getting an oyster's **pearl** unstuck as a metaphor for getting the cough out of your throat. It's a coincidence the benzonatate contains "b-e-n-z-o, benzo," it's not a benzodiazepine.

G, guaifenesin, is an expectorant pronounced as "g" plus "why" that helps move phlegm out of the body for a more productive cough. It loosens secretions in the lower respiratory tract and increases secretions in the upper respiratory tract. It can cause dose-related nausea. We see it alone or in combination products, especially with dextromethorphan. The "p-e-c-t-o-r, pector" in expectorant looks like pectoralis or chest muscles. Students also connect guaifenesin with the green Mr. Mucus from **Mucinex** commercials.

D, dextromethorphan, a cough suppressant. Note, dextromethorphan is a drug of abuse and, in high doses, causes euphoria and hallucinations. Despite abuse, there is little regulation. It's often available OTC. Dextromethorphan's brand **Robitussin** "robs" your cough and "tussin" resembles "tussive." Anti*tussives* are anti-cough medicines. Watch for dextromethorphan in combination cold products.

CO, codeine, an opioid cough suppressant, stronger than dextromethorphan, also pairs with guaifenesin. As a cough medicine, it is a DEA Schedule V drug to contrast the DEA Schedule II codeine tablet. Codeine causes respiratory depression in excess. Use that "cough" and "codeine" both start with "c-o" to connect them. The "chera" in guaifenesin with codeine's brand **Cheratussin AC** comes from the product's cherry flavoring. What the "AC" means is debatable; some students think "anti-cough" although it's probably "and codeine."

H, hydrocodone, a DEA Schedule II opioid, combines with chlorpheniramine, an antihistamine, to relieve a sore throat, fever, or body aches as **Tussionex**. Brand name Tussionex is a pineapple-flavored liquid anti*tussi*ve, hence the brand name.

If your cough gets worse, you may need the steroids in the next slide.

Question 36. Name four corticosteroids and any stems in the prefix, infix, or suffix position.

NOT PUMPED STEROIDS

P_____ - stem: _____

U

M_____ - stem: _____

P_____ - stem: _____

E

D_____

36. Oral Corticosteroids – NOT PUMPED STEROIDS

> **P**rednisone (Deltasone)
>
> **U**
>
> **M**ethyl**pred**nisolone (Medrol)
>
> **P**rednisolone (OraPred)
>
> **E**
>
> **D**examethasone (Decadron)

Quick Summary

Oral corticosteroids treat inflammation. While we see their use in acute exacerbations, they also work for other inflammatory conditions like rheumatoid arthritis, lupus, multiple sclerosis, and psoriasis.

We use the mnemonic NOT PUMPED STEROIDS to suggest these are therapeutic steroids, not drugs abused by some athletes. Also, these aren't nasal sprays.

P, prednisone, is an intermediate-acting steroid. To remember the side effects, some students associate prednisone with feeling like they are in the Twilight Zone. Hear the Twilight Zone theme song in your mind, and you'll be sure to remember.

M, methylprednisolone, is more potent and needs 4 mg to equal 5 mg of prednisone or prednisolone. We often see methylprednisolone as Medrol Dosepak, a package that spells out six-day tapering. Most students know **prednisone**, but many steroid compounds have this unofficial "sone" (s-o-n-e) ending. In prednisone, we find "pred, p-r-e-d," the official stem, as a *prefix*. In methyl**pred**nisolone, the stem is an *infix*.

P, prednisolone, is an intermediate-acting steroid often found in liquid preparations. Prednisolone's brand **OraPred** combines "ora, o-r-a" from "oral," its dosage form with the *pred* stem. Orapred.

D, dexamethasone, is long acting and the most potent steroid, as 0.75 mg equals 4 mg of methylprednisolone and 5 mg of prednisolone and prednisone.

From steroid drug names, we move to steroid side-effects.

RESPIRATORY MNEMONIC FLASHCARDS

Question 37. Name four side effects of chronic steroid use.

LONE

L_____ R_____

  Example 1. B_____ H_____

  Example 2. M_____

  Example 3. P_____

O_____ and I_____

N_____ C_____

  Example 1. M_____

  Example 2. I_____

E_____ G_____ and GI _____

141

37. STEROID SIDE EFFECTS, -LONE

> **L**ipid redistribution (Buffalo Hump, Moonface, Potbelly)
>
> **O**steoporosis and immunosuppression
>
> **N**euro changes such as mood and insomnia
>
> **E**levated glucose and GI effects (take with food)

QUICK SUMMARY

These steroid side-effects come from chronic use. To remember this list, we take the last four letters of methylprednisolone and prednisolone to make our –LONE mnemonic.

Early effects include insomnia and nausea. Patients can also experience weight gain and increased appetite, mood swings, and raised blood pressure, blood sugar, and restlessness.

Long term use causes bone weakening, poor wound healing, and increased infection risk, growth retardation, buffalo hump, irregular menstruation, and acne.

The concern is higher when patients take more than 20 mg of prednisone or prednisolone, or its equivalents daily for over two weeks. After extended therapy, prescribers should taper patients off steroids to allow the adrenal gland time to begin producing cortisol again.

Next, we next move to inhaled steroids.

RESPIRATORY MNEMONIC FLASHCARDS

Question 38. Name four inhaled corticosteroids and what patients should do after inhaling to prevent thrush.

FUMBLE TO BREATHE

F_____ (steroid)

U_____ (to avoid thrush)

M_____ (steroid)

B_____ (steroid)

L

E

B_____ (steroid)

38. INHALED CORTICOSTEROIDS – FUMBLE TO BREATHE

> **F**luticasone (Flovent)
>
> **U**se water to rinse and avoid thrush
>
> **M**ometasone (Asmanex)
>
> **B**udesonide (Pulmicort)
>
> **L**
>
> **E**
>
> **B**eclomethasone (Qvar)

QUICK SUMMARY

Asthmatics have airway constriction and inflammation. Asthma treatment includes rescue medications for respiratory symptoms like immediate shortness of breath. Maintenance inhaled corticosteroids like these reduce inflammation helping keep symptoms at bay

Note inhaled steroid brand names differ from intranasal drugs. Inhalers have slightly different delivery mechanisms, so inhalation counseling varies.

We categorize inhaled corticosteroids as low, medium, or high, and adjust depending on asthma severity. Potencies vary by inhalation number per day or medication dose per puff. Note, anti-inflammatories are inappropriate for *acute* asthma attacks.

Our mnemonic is FUMBLE TO BREATHE, something these anti-inflammatories help.

Fluticasone's "sone, s-o-n-e" ending is a useful clue that flutica*sone* is a steroid. In the past, inhalers felt cold when patients used them because the chlorofluorocarbon (C-F-C) propellant spray was similar to Freon, the refrigerant used in air conditioning. However, C-F-Cs damage the ozone layer, so the new ozone-safe hydrofluoroalkane (H-F-A) replaces the CFC propellant. The brand name **Flovent H-F-A** uses the first two letters of the generic name **fluticasone**, incorporates f-l-o from "airflow," and adds "vent" for ventilation. The diskus, d-i-s-k-u-s inhaler is a device that looks like an Olympic discus, d-i-s-c-u-s. It employs a dry powder for inhalation rather than a propellant and liquid.

U, use water to rinse and avoid thrush. Inhaled corticosteroids can cause thrush, a fungal mouth infection. Patients should rinse their mouth out with water after each inhaler use to avoid infection.

M, mometasone's brand **Asmanex** is "asthma" plus "ex" – meaning "get rid of asthma."

B, budesonide's brand **Pulmicort** is pulmonary, which means lungs, plus "cort" short for corticosteroid.

B, beclomethasone, brand Qvar, is another corticosteroid. The brand name might use the a-r from "air."

The next slide also helps with asthma and COPD, but these medications have a different mechanism of action that helps with bronchodilation.

Memorizing Pharmacology Mnemonics

Question 39. Name four inhaled beta-2 agonist bronchodilators from short to ultra-long-acting and identify the class suffix.

ALL SAVE LUNGS

A_____ (short-acting)

L_____ (short-acting)

L

S_____ (long-acting)

A

V_____ (extra-long acting)

E

Class suffix: -t_____

39. Beta-2 Agonists – ALL SAVE LUNGS

> **A**lbuterol (ProAir HFA, Proventil HFA, Ventolin HFA)
> **L**evalbuterol (Xopenex HFA)
> **L**
>
> **S**almeterol (Serevent Diskus)
> **A**
> **V**ilanterol (in combination)
> **E**

Quick Summary

Beta-2 agonists are sympathomimetic agents that work directly on the lungs to cause bronchodilation. We use them in asthma and COPD. ALL SAVE LUNGS is our mnemonic.

A, albuterol is a rescue inhaler for asthma and COPD. Patients can use it every four to six hours as needed. However, if patients need many doses daily, we often try to use prophylactic medicines. While beta-2 agonists are selective for beta-2 receptors in the lungs, increased doses from overuse might lead to selectivity loss, where you see activation of beta-1 receptors in the heart causing tremor, palpitations, and tachycardia. There are also beta-2 receptors in the myocardial tissue which causes tachycardia without prolonged exposure or overuse.

Albuterol's "terol, t-e-r-o-l" stem suggests it's a bronchodilating beta-2 adrenergic agonist. Albuterol's "terol" stem *does not* help you distinguish long-acting *or* short-acting. Memorize the distinction. The brand name **ProAir HFA** is straightforward with "Pro" as in "I'm for it" and "Air" for airway.

L, levalbuterol, is albuterol's R-isomer which might have less effect on heart rate and cause fewer tremors at high doses. Think of "open" and "exhale" with levalbuterol's brand **Xopenex**.

S, salmeterol, is a long-acting beta agonist for COPD maintenance. We don't use this as monotherapy in asthma because of the risk of asthma-related death. We need the patient to lack good control with an inhaled corticosteroid before adding this on. Meter, as in metered dose inhaler, is in the middle of sal*meter*ol.

V, vilanterol is an ultra-long-acting beta-2 agonist only available in combination with inhaled corticosteroids or long-acting muscarinic antagonists.

Formoterol is another essential beta-2 agonist we'll address in the combination products.

Side note: Outside the U.S., levalbuterol is levosalbutamol and albuterol is salbutamol.

Next, we discuss other beta and alpha receptor medications for context.

Question 40. Provide the effect of an alpha-1, beta-1, and beta-2 agonist and antagonist. Give a medication example of all six possibilities.

ALPHA BETA BOX

Alpha-1 agonist effect: Vaso_____

    Medication example: P_____

Alpha-1 antagonist effect: Vaso_____

    Medication example: D_____

Beta-1 agonist effect: Inc_____

    Medication example: D_____

Beta-1 antagonist effect: Dec_____

    Medication example: M_____

Beta-2 agonist effect: Broncho_____

    Medication example: A_____

Beta-2 antagonist effect: Broncho_____

    Medication example: P_____

40. ALPHA AND BETA AGONISM AND ANTAGONISM – ALPHA BETA BOX

	Agonist	Antagonist
Alpha₁	Vasoconstriction (increases BP) Phentolamine (OraVerse)	Vasodilation (decreases BP) Dox<u>azosin</u> (Cardura XL)
Beta₁ (heart)	Increases heart rate Dobutamine (Dobutrex)	Decreases heart rate Metop<u>rolol</u> (Toprol XL)
Beta₂ (lungs)	Bronchodilation (increase airflow) Albu<u>terol</u> (ProAir HFA)	Bronchoconstriction (decrease airflow) Propran<u>olol</u> (Inderal)

QUICK SUMMARY

The ALPHA BETA BOX puts alpha and beta agonists and antagonists in an easy-to-digest three-by-three box. While the alpha-1 and beta-1 drugs mostly work in the cardio section, the beta-2 agonists work here in the respiratory chapter. Let's go over some bronchodilating beta-2 agonists as part of a bronchodilator and steroid combination in the next slide.

Question 41. Provide four beta-2 agonist/corticosteroid combination products and the beta-2 agonist class suffix.

7 F's PROPEL BREATH MOVES

F_____ F_____ /B_____

F_____ F_____ /M_____

F_____ F_____ /V_____

F_____ P_____ /S_____

Beta-2 agonist class suffix: -t_____

41. BETA-2 AGONISTS + INHALED CORTICOSTEROIDS – 7 Fs PROPEL BREATH MOVES

> **F**ormo<u>terol</u> fumarate / **B**udesonide (Symbicort)
>
> **F**ormo<u>terol</u> furoate / **M**ometasone (Dulera)
>
> **F**luticasone furoate / **V**ilan<u>terol</u> (Breo Ellipta)
>
> **F**luticasone **pro**pionate / **S**alme<u>terol</u> (Advair)

QUICK SUMMARY

Combination products reduce inflammation with corticosteroids and cause bronchodilation with beta-2 agonists. Since these inhalers contain inhaled corticosteroids, patients should rinse their mouth out after use to prevent thrush. Our mnemonic 7 Fs PROPEL BREATH MOVES reminds us of these combinations importance in respiratory disorders.

FF, B, formo<u>terol</u> fumarate / budesonide, is a metered-dose inhaler for both COPD and asthma. The "terol, t-e-r-o-l" stem in **formoterol** marks a beta-2 agonist bronchodilator. Similar to **fluticasone** paired with **salmeterol** in **Advair, budesonide** has the "sone, s-o-n-e" syllable. We pronounce it "sone," though it's spelled "s-o-n." You can think of the "S-y-m" in the brand name **Symbicort** as symbiosis, meaning "working with," plus "c-o-r-t" for corticosteroid: Sym-bi-cort.

FF, M, formo<u>terol</u> / mometasone, is a metered-dose inhaler for asthma only. The formoterol furoate / mometasone combination's brand **Dulera** combines "dual," meaning two, and the sound from "air" with e-r-a, era instead of a-i-r, air. "Dual air" for trouble breathing. Dulera.

FF, V, fluticasone furoate / vilanterol, has a unique delivery device and is for COPD and asthma. The "Ellipta" in **Breo Ellipta** refers to the ellipse shape, which is really a stretched circle that matches the Ellipta inhaler's shape. In Washington D.C., the Ellipse is the public park between the White House's backyard and the Washington Monument.

F, PRO, S, fluticasone propionate and salmeterol. This Advair Diskus is a dry powder inhaler for asthma and COPD, while the metered-dose inhaler is only for asthma. Recognize the steroid fluticasone by the unofficial "sone," and the long-acting bronchodilator salmeterol by the "terol" stem. The brand name **Advair** seems like "add two drugs to get air." Ad-vair.

Beta-2 agonists bronchodilate, but so do muscarinic receptor antagonists, which we'll move to next.

Question 42. Name five anticholinergics for respiratory disorders, a therapeutic use, their delivery mode, and a possible side effect.

ANTIMUSCARINIC

A_____ (prototype anticholinergic)

N

T_____ (long-acting anticholinergic)

I_____ (short-acting anticholinergic)

M

U_____ (long-acting anticholinergic)

S

C_____ (therapeutic use or condition)

A_____ (long-acting anticholinergic)

R

I_____/A_____ (short-acting anticholinergic / short-acting beta2 agonist)

N

I_____ (mode of delivery)

C_____ (possible side effect)

42. Muscarinic Antagonists – ANTIMUSCARINIC

Anticholinergic / atropine
N
Tio<u>tropium</u> (Spiriva)
Ipra<u>tropium</u> (Atrovent)

M
Ume<u>clidinium</u> (Incruse Ellipta)
S
COPD
A<u>clidinium</u> (Tudorza)
R
Ipra<u>tropium</u> / Albu<u>terol</u> (DuoNeb, Combivent)
N
Inhaled
Cough

QUICK SUMMARY

Muscarinic antagonists, also known as ANTIMUSCARINICs or anticholinergics, block bronchial constriction caused by acetylcholine. We can use them alone or combined with beta-2 agonists for asthma and COPD. We use combination products later in disease progression.

Antimuscarinics can produce dry mouth, but systemic side effects like urinary retention and constipation are rare with these inhaled drugs. Other side effects include a cough and bitter taste. ANTIMUSCARINIC serves as our mnemonic.

A, anticholinergic / atropine. Instructors use **atropine** as the prototype drug for the anticholinergics, and you can see the "trop, t-r-o-p" stem in **atropine, ipratropium**, and **tiotropium**.

T, tiotropium, is a long-acting anticholinergic with a device that punctures capsules to release an inhaled dry powder. Counsel patients *not to swallow* the capsules. Tiotropium has the same "trop, t-r-o-p" stem as ipratropium and is the long-acting version. As with the "terols," the beta-2 receptor agonists, you have to memorize which anticholinergic is long-acting versus short-acting. The brand name **Spiriva** takes the "spir, s-p-i-r" from respire, like respiration.

I, ipratropium, is a short-acting anticholinergic available alone or combined with albuterol as inhaler or nebulizing solution. In brand name **Combivent**, we see "combination" and "ventilation." Another brand name, **DuoNeb**, suggests a drug duo in nebulized form; Duo-Neb.

U, umeclidinium, a long-acting antimuscarinic, is available alone or in combination with vilanterol. **Umeclidinium** and **aclidinium's** –clidinium stem represents muscarinic receptor antagonists. Think of umeclidinium's Incruse, I-n-c-r-u-s-e Ellipta as increase, i-n-c-r-e-a-s-e air to the patient.

COPD, is chronic obstructive pulmonary disease and includes emphysema and chronic bronchitis. We often treat a COPD patient differently than an asthma patient.

A, ac**lidinium**, is a long-acting anticholinergic.

I, inhaled, reminds us of the dosage form.

C, cough, a possible side-effect.

Now we move on to respiratory drugs that don't neatly fit into a single category.

Memorizing Pharmacology Mnemonics

Question 43. Name two leukotriene inhibitors, their class suffix, a monoclonal antibody IgE inhibitor for severe allergy, the monoclonal antibody class suffix, and an anaphylaxis medication.

Z'MORE RESPIRATORY DRUGS

Z_____ (leukotriene inhibitor)

M_____ (leukotriene inhibitor)

O_____ (monoclonal antibody)

Respiratory

E_____ (anaphylactic medication)

Leukotriene inhibitor class suffix: -l_____

Monoclonal antibody class suffix: -m_____

43. Leukotriene inhibitors, Anti-IgE, Epi – Z'MORE RESPIRATORY DRUGS

> **Z**afir<u>lukast</u> (Accolate)
>
> **M**onte<u>lukast</u> (Singulair)
>
> **O**mal<u>izumab</u> (Xolair)
>
> **R**espiratory
>
> **E**pinephrine (EpiPen)

Quick summary

The last mnemonic encompasses a hodgepodge of respiratory agents we combine as Z'MORE RESPIRATORY DRUGS.

Z, zafir<u>lukast</u>, inhibits the leukotrienes that form in leukocytes, white blood cells, and cause inflammation. By blocking them, zafirlukast helps asthma inflammation. The "-lukast, l-u-k-a-s-t" stem comes from leukotriene.

M, monte<u>lukast</u>, also inhibit leukotrienes responsible for inflammation. We can add leukotriene inhibitors at many stages in asthma treatment. Common side effects include headache and dizziness. Montelukast's brand **Singulair** comes from "single" daily dosing and "air" for asthmatic lungs – Singulair.

O, omal<u>izumab</u> is a monoclonal antibody that inhibits IgE binding in allergy and severe asthma. A healthcare professional must inject it subcutaneously every two to four weeks because of anaphylaxis risk, especially after the first dose. Like infliximab for ulcerative colitis, omalizumab is a biologic.

The "oma-" (o-m-a) prefix separates it from other similar drugs.

The "-li-" (l-i) stands for immunomodulator (the target),

The "-zu-" (z-u) stands for humanized (the source),

And the "-mab" (m-a-b) is for monoclonal antibody.

The anaphylaxis black box warning pertains to the first dose and even a year after treatment onset.

Therefore, health providers inject omalizumab where a medicine for treating anaphylaxis is available. Omalizumab's brand name **Xolair** seems like "exhale" and "air."

R, respiratory, reminds us of the drugs' therapeutic use.

E, epinephrine, is a sympathomimetic agent for anaphylaxis.

The word **epinephrine** has a Greek origin. "Epi" means "above," and "neph" means "kidney." Above the kidney is the adrenal gland responsible for the body's natural release of **epinephrine**. The Latin version of **epinephrine** is adrenaline [[uh dren uh Lynn, not uh dreen uh Lynn or uh dren uh leen]]. The "ad" means "above," and "renal" means "kidney" – adrenaline. Either version stimulates the fight or flight response that can help a patient in an anaphylactic state. The **EpiPen** brand name comes from the pen-like auto-injector.

That wraps up the R, respiratory section. Let's move on to I, Immune in our GM-RINCE mnemonic.

CHAPTER 4: IMMUNE MNEMONIC FLASHCARDS

Question 44. Name the penicillin class suffix, two penicillin dosage forms, one PO and one IM, and name two amino penicillins in order of preference for PO indications.

TWO PENICILLINS AND TWO AMINOPENICILLINS

P_____ (_ _) PO form

P_____ (__) IM form

A_____ (most preferred for PO dosage form)

A_____ (least preferred for PO dosage form)

Penicillin class suffix: -c_____

44. Penicillins – TWO PENICILLINS AND TWO AMINOPENICILLINS

Peni<u>cillin</u> VK, PO (Veetids)
Peni<u>cillin</u> G, IM (Bicillin)

Amoxi<u>cillin</u> (Moxatag)
Ampi<u>cillin</u> (Principen)

Quick Summary

We begin with cell wall synthesis inhibitors, or beta-lactams, which include penicillin, cephalosporin, carbapenem, and monobactam antibiotics. Penicillins bind penicillin binding proteins within bacterial cell membranes, inhibiting crosslinking.

Penicillins cause GI effects like nausea, diarrhea, and hypersensitivity reactions. These range from a mild rash to life threatening anaphylaxis. True penicillin allergies are rare. Patients report GI upset or a black hairy tongue mistakenly as allergic symptoms. However, the possibility of life-threatening anaphylaxis often moves treatment to other drug classes. On a trial later in life, most patients are no longer allergic, but we understandably don't take the chance.

Penicillins display time-dependent killing, which means the drug will be more effective the longer its concentration remains above a minimum inhibitory concentration (MIC). They are bactericidal, killing bacteria, rather than bacteriostatic, which inhibits growth.

To remember drugs with the same words or letters, it's easier to put two pair together as 2 PENICILLINS and 2 AMINOPENICILLINS.

2 penicillins

Penicillin G is a parenteral formula dosed in million-units.

Penicillin VK is an oral, enteral formula absorbed better on an empty stomach. Penicillins cover Gram-positive organisms and are first-line treatments for streptococci. The "cillin" stem sounds like "*cell*-in" to help you remember penicillins destroy the *cell* wall.

2 aminopenicillins

Amoxicillin is only PO and has excellent absorption compared to other beta lactams. Patients take it with or without food in multiple daily doses. Side effects are similar to penicillins. We prefer amoxicillin over ampicillin because of fewer GI effects and fewer daily doses.

Amoxicillin has the penicillin class "cillin, c-i-l-l-i-n" stem. The "a-m-o" shows it's an "amino" penicillin.

The brand **Moxatag** mixes the first four letters of generic amoxicillin.

Ampicillin, comes PO, by mouth, or IV. It has better intraabdominal penetration than amoxicillin for GI infections outside of the GI tract.

Like penicillins, aminopenicillins cover streptococcus but not staphylococcus, but we can use them to cover *enterococcus faecalis*. Like penicillins, aminopenicillins show bactericidal, time-dependent killing.

When bacteria resists penicillin, you can attack with penicillinase resistant penicillins, which we'll cover next.

Question 45. Name a combination penicillin with beta-lactamase resistant medication, what clavulanate gives to amoxicillin, and three other beta-lactamase resistant medications.

LACTAM (ACED) NO MORE

A_____/C_____ (penicillin antibiotic / beta-lactamase inhibitor)

E_____ (what clavulanate gives amoxicillin)

D_____ (penicillin for *staphylococcus*)

N_____ (beta-lactamase resistant penicillin)

O_____ (beta-lactamase resistant penicillin)

M_____ (beta-lactamase resistant penicillin)

ORE

45. Penicillinase Resistant Penicillins – Lactam (ACED) No More

> **A**moxicillin /
> **C**lavulanate (Augmentin)
> **E**xtra Beta-Lactamase Coverage
> **D**icloxacillin (Dynapen)
>
> **N**afcillin (Unipen)
> **O**xacillin (Bactocil)
>
> **M**ethicillin
> ORE

Quick Summary

LACTAM(ACED) should end with –ase, a-s-e but ACED, A-C-E-D allowed me to put

A, amoxicillin,

C, clavulanate, and

E, extra beta-lactamase coverage with dicloxacillin below. Amoxicillin and ampicillin come as combination products with beta-lactamase inhibitors.

A beta-lactam ring is a chemical part of penicillin and a beta-lactamase is an enzyme bully trying to break it. Adding clavulanate is like giving amoxicillin a bodyguard to protect the beta lactam ring. Adding a beta lactamase inhibitor provides coverage against anaerobes and organisms that make this enzyme, like *Haemophilus influenzae*. We often see amoxicillin with clavulanate, to treat resistant otitis media, a middle ear infection. The clavulanate often causes diarrhea, so prescribers use the lowest possible dose.

When **amoxicillin** alone doesn't work, we can think of the brand **Augmentin**, as it augments **amoxicillin's** defenses against the bacterial beta-lactamase enzyme with **clavulanate**.

I think of the "clavicle," the bone in your shoulder, as being protective of the upper lung and associate **clavulanate** with that same protective effect since both start with c-l-a-v.

D, dicloxacillin is the only oral anti-staph penicillin. We reflect this dynamic coverage of staph with a penicillin with dicloxacillin's brand name **Dynapen**, a dynamic penicillin

NO MORE

The penicillinase-resistant penicillins, **nafcillin**, **oxacillin** and **methicillin** are resistant on their own to penicillinases. The following three are basically the same thing – methicillin refers to MRSA or MSSA, oxacillin is for testing, and nafcillin is the prescribed drug.

N, nafcillin, is available IV. Brand **Unipen** reminds us this group is penicillinase resistant with a single, uni, penicillin, unlike Augmentin which requires two drugs.

O, oxa<u>cillin</u>, is available IV and PO. The penicillinase-resistant penicillins cover methicillin-sensitive *Staphylococcus aureus* (MSSA) infections.

M, methi<u>cillin</u>, we don't see anymore because it causes renal failure, but we use 'methicillin' in methicillin-resistant *Staph aureus* (MRSA) to express that we have a resistant bug, so I've included it here.

Side effects follow those of penicillins and aminopenicillins. The body metabolizes nafcillin and oxacillin hepatically, so we don't need to adjust for renal insufficiency. Like the other beta lactams, the penicillinase-resistant penicillins display time-dependent killing and are bactericidal.

Penicillin allergic patients have a potential cross allergic reaction to our next group, cephalosporins.

Memorizing Pharmacology Mnemonics

Question 46. Name the two cephalosporin class prefixes, write nine cephalosporins in generational order, and outline the three general trends regarding Gram coverage, resistance to beta-lactamases, and CSF penetration as you move up the generations.

1) CEPH ONE 2) OR OX 3) TRI DIN TAZ 4) PIME 5) TAROL

Cephalosporin class prefixes: _____ & _____

Cephalosporins (in generational order)
1st gen:
CEPH, Ceph_____ (1st)
ONE, Cef_____ (1st)

2nd gen:
OR, C_____ (2nd)
OX, C_____ (2nd)

3rd gen:
TRI, C_____ (3rd)
DIN, C_____ (3rd)
TAZ, C_____ (3rd)

4th gen:
PIME, C_____ (4th)

5th gen:
TAROL, C_____ (5th)

Newer generation cephalosporins tend to be _____ in terms of Gram coverage, resistance to beta-lactamases and penetration into the CSF compared to older ones.

46. CEPHALOSPORIN GENERATIONS – 1) CEPH ONE 2) OR OX 3) TRI DIN TAZ 4) PIME 5) TAROL

> *CEPH*, **C**ephalexin (Keflex, 1st)
> *ONE*, **C**efazolin (Kefzol, 1st)
>
> *OR*, **C**efaclor (Ceclor, 2nd)
> *OX*, **C**efuroxime (Ceftin, 2nd)
>
> *TRI*, **C**eftriaxone (Rocephin, 3rd)
> *DIN*, **C**efdinir (Omnicef, 3rd)
> *TAZ*, **C**eftazidime (Fortaz, Tazicef, 3rd)
>
> *PIME*, **C**efepime (Maxipime, 4th)
> *TAROL*, **C**eftaroline (Teflaro, 5th)

QUICK SUMMARY

The next beta-lactams we'll cover are five cephalosporin generations. You'll notice the prefixes are the same: either ceph-, c-e-p-h or cef-, c-e-f. Newer generation advantages usually include better:

1) Gram-negative coverage, and

2) Resistance to beta-lactamases.

3) Penetration into the cerebrospinal fluid (C-S-F)

G, Gram-negative coverage, R, resistance, and P, penetration into the CSF as in "get a GRIP, G-R-I-P" on cephalosporin generations helps us remember. It may be easier to remember that generations one and five have excellent Gram-positive coverage, and Gram-negative coverage increases with generations with a peak at generation four.

General cephalosporin side effects match penicillins, including GI upset and diarrhea. There is a small cross-sensitivity risk. We therefore avoid cephalosporins with a penicillin allergic patient; otherwise, we risk angioedema or anaphylaxis.

Cephalosporins, like penicillins, display bactericidal time-dependent killing. To recognize which generation a drug falls into, I made meaning from meaningless drug name parts.

CEPH, cephalexin, is a first-generation cephalosporin with a narrow activity spectrum covering Gram-positive streptococci and staphylococci. We use these for skin infections and strep throat. First generation cephalosporins have the worst Gram-negative activity, CNS penetration, and beta-lactamase resistance.

I believe people confused the medical prefix ceph, c-e-p-h meaning "head" with this drug prefix, so they switched it to cef, c-e-f for later generations. You have "one" head, ceph-, use that to remember "first generation." of activity. The brand name **Keflex** begins with a "K" and takes some letters from c<u>e</u>phal<u>ex</u>in.

ONE, cefazolin is also a first-generation cephalosporin and you can find an o-n-e in the word cefazolin to remember.

OR, cefaclor. We can take the "or, o-r" as *two* letters to remember second-generation. It covers the same bugs as first-generation agents, adding *Haemophilus* and *Neisseria*. We use them for otitis media and pneumonia. In cefaclor, be careful, as this doesn't mean all drugs that end in "or, o-r" are second-generation cephalosporins – it's just a way to put cefaclor in its generation. **Cefuroxime** has a two-letter o-x, ox you can use to remember second generation.

TRI, ceftriaxone, and other third generation cephalosporins confer more activity against Gram-negative organisms. Ceftriaxone is available IV and IM, and has convenient once or twice daily dosing. The "tri, t-r-i" from **ceftriaxone** means "three," like a *tri*cycle with three wheels. **Rocephin**, the brand name, comes from Hoffman-LaRoche's patent. The drug company took the "Ro" (r-o) from "LaRoche," and the "ceph" (c-e-p-h) and "in" from cephalosporin to make **Ro-ceph-in**.

DIN, cefdinir, I have personal experience with. My daughter had a double ear infection while taking amoxicillin. Since that didn't work, we needed a beta-lactamase resistant, third-generation cephalosporin like cefdinir to clear the infection. It's one of the only cephalosporins you don't refrigerate after reconstitution. With **cefdinir**, I looked at the "din, d-i-n," which means noisy. My triplets are noisy, and one daughter needed cefdinir for a resistant ear infection. This helped me connect cefdinir and third generation.

TAZ, ceftazidime, is IV, and important because it covers pseudomonas, a multidrug resistant Gram-negative rod. It's available with avibactam, a beta lactamase inhibitor. We often use third-generation agents for pneumonia, meningitis, and pyelonephritis. The "taz, t-a-z" also has three memorable letters in **ceftazidime**. Memorize three at a time in the third generation – ceftriaxone, ceftazidime, and cefdinir to further strengthen the connection.

PIME, cefepime, is a fourth-generation cephalosporin that covers Gram-negative organisms plus pseudomonas. Cefepime comes IV or IM and has some Gram-positive activity. Remember fourth generation with the four letters – "p-i-m-e" – in cefepime and brand **Maxipime**. When released, Maxipime *was* the *max*imum generation.

TAROL, ceftaroline, is an IV fifth-generation cephalosporin with Gram-positive coverage similar to that of the third generation. It does not cover pseudomonas, but it covers MRSA and other multidrug resistant organisms. Take the *five* letter *tarol*, t-a-r-o-l from **ceftaroline**, to remember fifth generation.

Next, we'll review the monobactam aztreonam, structurally related to ceftazidime.

Question 47. Name a monobactam that we can use if a patient is penicillin allergic.

Monobactam for patients allergic to penicillin

A_____

47. Monobactam – AZTREONAM ALONE

> *A*ztre<u>onam</u> (Azactam, Cayston) *ALONE*

Quick Summary

Aztreonam is the only monobactam available, therefore AZTREONAM ALONE is our mnemonic. Like previous beta-lactams, it inhibits cell wall synthesis. It only covers Gram-negative bacteria. Aztreonam's value comes from it *not* having cross-reactivity with penicillins and other beta lactams for penicillin allergic patients.

Aztre<u>onam</u> has similar side effects to penicillin, including GI upset and rash. Like other beta lactams, we see bactericidal time-dependent killing.

When I looked up the monobactam suffix, I saw monam, m-o-n-a-m, rather than –onam, o-n-a-m. The vowels in the drug name's center dictated this. Brand name **Azactam** takes the first two letters of **aztreonam** and combines it with the last four letters of monobactam, its drug class.

Moving on, we go to carbapenems.

Question 48. Name four carbapenems, the class suffix, and indicate which drug is no longer available.

MEDIC

M_____ (carbapenem)

E_____ (carbapenem)

D_____ (carbapenem no longer available)

I_____ (carbapenem)

C_____ (medication class)

Carbapenems class suffix: -p_____

48. Carbapenems – MEDIC

> **M**eropenem (Merrem)
>
> **E**rtapenem (Invanz)
>
> **D**oripenem (Doribax) [no longer available]
>
> **I**mipenem (Primaxin IV)
>
> **C**arbapenems –penem stem

Quick Summary

The MEDIC mnemonic holds carbapenems, our last beta-lactam antibiotics drug class that inhibits cell wall synthesis. Carbapenems are intravenous, IV, and broad spectrum. They cover Gram-positive and Gram-negative organisms and anaerobes. They also fight extended-spectrum beta-lactamase producing organisms not including MRSA or VRE. Meropenem, doripenem, and imipenem all work against pseudomonas, but ertapenem does not. Carbapenems treat polymicrobial infections or work as empiric therapy for resistant organisms.

All carbapenems can lower a seizure threshold, as they cross the blood brain barrier and enter the CNS especially at high doses. Imipenem is the worst and has the highest seizure risk. Therefore, use it cautiously in patients with seizure disorders. Carbapenems also decrease levels of the antiepileptic valproic acid, which can lead to seizures. We avoid carbapenems in penicillin-allergic patients because of their similarity. This is controversial. Practitioners might use a carbapenem rather than a 4th generation cephalosporin. The cross-reactivity is low.

Carbapenems display bactericidal time-dependent killing.

The **carbapenems** have the **–penem, p-e-n-e-m** suffix. Most brand names take the first two or three letters of the generic name, so we get brand *Mer*rem from **mero*penem***, **Dori**bax from *dori*****penem***, and **Primaxin IV** from ***imi*penem**. We also see imipenem combined with cilastin to help avoid elimination.

Next, we visit vancomycin, a cell wall synthesis inhibitor active against MRSA.

Memorizing Pharmacology Mnemonics

Question 49. Name a glycopeptide, a severe toxicity associated with rapid infusion, and five severe therapeutic indications with a trough of 15 to 20 mcg/mL.

VANCOMYCIN RED HOMES

V_____ (glycopeptide drug)

Red ____ _____ (severe toxicity with rapid infusion)

Five therapeutic indications with a trough of 15-20 mcg/mL

H_____

O_____

M_____ (or "M" for covers MRSA)

E_____

S_____

49. Glycopeptide – VANCOMYCIN RED HOMES

> **V**anco<u>myc</u>in (Vancocin)
>
> **R**ED man syndrome
>
> **H**ospital-associated pneumonia
> **O**steomyelitis
> **M**eningitis (or M for "covers MRSA")
> **E**ndocarditis
> **S**epsis

Quick Summary

Like penicillins, glycopeptide antibiotics like **vancomycin** inhibit cell wall synthesis, binding to the D-alanyl-D-alanine part of the cell wall. This, in turn, blocks peptidoglycan polymerization. Vancomycin has zero oral absorption, so practitioners give vancomycin IV for systemic infections and orally for *Clostridium difficile* because it just needs to get to the GI tract. If you can't find vancomycin, it's in the refrigerator.

RED. Infusing vancomycin too fast can cause hypotension, flushing, and chills as part of red man syndrome. Most hospital protocols require infusion over at least one hour, usually no faster than 500 mg / 30 min for higher doses. Vancomycin is nephrotoxic, so we monitor renal function, and patients may rarely experience a severe rash. Vancomycin displays bactericidal, time-dependent killing.

When dosing vancomycin, we aim for trough concentrations at a steady state to ensure efficacy. We measure this by looking at the ratio of AUC/MIC (Area Under the Curve/Minimum Inhibitory Concentration). These trough concentrations are lower (10-15 mcg/mL) for milder infections, and higher (15-20 mcg/mL) for severe infections like pneumonia, meningitis, and endocarditis.

The HOMES acronym reminds us of infections vancomycin works against at a trough of 15-20:

H, hospital-associated pneumonia

O, osteomyelitis

M, meningitis

E, endocarditis

S, sepsis

Vancomycin only covers Gram-positive infections, most notably MRSA. Vancomycin is unique in that we measure its efficacy by comparing the AUC/MIC ratio. Note that vancomycin is for cellulitis and *C. Diff*, as well, which aren't part of the acronym.

Vanco<u>myc</u>in's "mycin, m-y-c-i-n" stem isn't useful for finding its therapeutic class. All "mycin" means is that chemists derived **vancomycin** from the *Strepto<u>myc</u>es* bacteria, taking the m-y-c from *Streptomyces*. Remember the coverage as "**vancomycin** will <u>van</u>quish MRSA." To remember the brand name **Vancocin**, remove the "my" from **vancomycin** to get V-a-n-c-o-c-i-n. **Vancocin**.

The next slide includes drugs similar to **vancomycin**, called *lipo*glycopeptides. All have Gram-positive coverage, as well.

Memorizing Pharmacology Mnemonics

Question 50. Name three lipoglycopeptides and the class suffix.

TRIPOD

T_____ (lipoglycopeptide)

RED _____ _____ (side effect of rapid administration)

I_____ (route of administration)

P_____ C_____ - _____ (mechanism of action)

O_____ (lipoglycopeptide)

D_____ (lipoglycopeptide)

Lipoglycopeptide class suffix: -v_____

50. Vancomycin related lipoglycopeptides – TRIPOD

> *T*ela<u>vancin</u> (Vibativ)
>
> *R*ed man syndrome
>
> *I*ntravenous administration
>
> *P*eptidoglycan cross-linking
>
> *O*rita<u>vancin</u> (Orbactiv)
>
> *D*alba<u>vancin</u> (Dalvance)

Quick Summary

I thought of the mnemonic TRIPOD since there's three lipoglycopeptides.

T, tela<u>vancin</u>, we dose once daily. It's teratogenic, so women of childbearing age need a negative pregnancy test. Telavancin interferes with INR and PT, two anticoagulant metrics for warfarin and aPTT, a heparin metric. However, it doesn't increase bleeding risk. The manufacturer doesn't recommend telavancin with nephrotoxic agents. The drugs' "**vancin, v-a-n-c-i-n**" stem helps you recognize the lipoglycopeptide antibiotic class. Telavancin's brand **Vibativ** includes Vi, v-i for victory against the resistant organisms, the b-a-t, bat from bacteria, and i-v, from its form of administration, IV.

R, red man syndrome, similar to vancomycin's effect, all three can cause the condition.

I, intravenous administration. We give all three lipoglycopeptides IV. They have similar coverage to vancomycin for MRSA and soft tissue skin infections.

P, polymerization. Lipoglycopeptides are also beta lactam antibiotics, which inhibit cell wall synthesis. They bind to the same cell wall part to block polymerization and the cross-linking of peptidoglycan to disrupt bacterial membrane potential. This allows lipoglycopeptides to show bactericidal concentration-dependent killing.

O, oritavancin, are both single dose regimens that have long half-lives. Oritavancin can cause falsely elevated PT, INR, and aPTT. As a CYP 2C9 inhibitor, it can increase bleeding risk with warfarin.

D, dalbavancin has a two-dose regimen, with the second dose following the first dose a week later. The manufacturers took the first generic name letters to make the brand name. Generic **oritavancin** becomes **Orbactiv** and **dalbavancin** becomes **Dalvance**.

Next, we'll cover concentration dependent antibiotics as a group.

IMMUNE MNEMONIC FLASHCARDS

Question 51. Name a combination of four antibiotic drug classes or drug names that have concentration dependent killing, what concentration dependent killing relies on, and the relative size of the doses versus frequency.

FRAMED CONCENTRATION CHART

F_____ (drug class)

R_____ of max concentration to MIC

A_____ (drug class)

M_____ (drug, nitroimidazole)

E_____ large doses less frequently

D_____ (drug, lipopeptide)

187

51. Time vs. Concentration Dependent – Framed Concentration Chart

> **F**luoroquinolones
>
> **R**atio of max concentration to MIC
>
> **A**minoglycosides
>
> **M**etro<u>nidazole</u> (Flagyl)
>
> **E**xtra-large doses less frequently
>
> **D**apto<u>mycin</u> (Cubicin)

Quick Summary

Antibiotics can exhibit time-dependent or concentration-dependent killing.

Time-dependent antibiotics we dose frequently. These include penicillins, cephalosporins, aztreonam, carbapenems, and vancomycin. Time-dependent killing relies on how long the drug's blood concentration stays above the minimum inhibitory concentration (MIC).

Concentration-dependent killing relies on a max concentration to MIC ratio. The goal is to keep a high peak for more killing and low troughs to avoid toxicity. We might give these drugs in larger doses, less often. The mnemonic FRAMED CONCENTRATION CHART alludes to the chart we see when verifying the concentration.

F, fluoroquinolones,

R, ratio of the max concentration compared to the MIC.

A, aminoglycosides

M, metronidazole, causes DNA strand breakage, inhibiting protein synthesis. It covers anaerobes, protozoal infections, *Clostridium difficile*, peptic ulcer disease, and intraabdominal infections. Metronidazole causes a disulfiram-like reaction when mixed with alcohol, resulting in nausea, vomiting, and cramping.

Gastroenterologists use **metronidazole** for peptic ulcer disease (P-U-D). **Metronidazole** is technically an antiprotozoal. Look at the "azole" (a-z-o-l-e) ending not as a stem, but for matching "ozoal" (o-z-o-a-l) from "protozoal."

Another student of mine learned to give "Flag" [[pronounce fladge]] a shorter form of metronidazole's brand **Flagyl**, for *B. frag* [[pronounce Bee Frag]], a shortening of the Bacteroides *fragilis* infections.

E, extra-large doses rather than more frequent doses.

D, daptomycin, is a cyclic lipopeptide and cell membrane inhibitor, not a cell wall inhibitor. We use daptomycin for MRSA, VRE, blood stream infections, endocarditis, and skin infections. Lung surfactant inactivates daptomycin, so it's ineffective for pneumonia. Daptomycin can increase myopathy and rhabdomyolysis risk. Suspend in patients with muscle pain or asymptomatic patients with an elevated CPK.

Side note:

Secnidazole (Solosec) treats bacterial vaginosis in women. Single dose treatment is effective; compared to metronidazole (Flagyl) and tinidazole (Tindamax), it doesn't influence the aldehyde dehydrogenase like metronidazole, making it a potentially better choice.

Next, we'll cover fluoroquinolones with chiefly Gram-negative coverage.

Question 52. Name four fluoroquinolones, the class suffix, and three considerations with regards to either side effects or drug interactions.

MEDICAL

M_____ (fluoroquinolone)

E_____ (caution)

D_____ (fluoroquinolone)

I_____ (drug interaction)

C_____ (fluoroquinolone)

A_____ (side effect)

L_____ (fluoroquinolone)

Class suffix –f_____

52. FLUOROQUINOLONES – MEDICAL

> **M**oxifloxacin (Avelox)
> **E**xtra careful with sunlight, as there's a risk of phototoxicity
> **D**elafloxacin (Baxdela)
> **I**nteracts with cationic multivalent antacids through chelation
> **C**iprofloxacin (Cipro)
> **A**chilles tendon pain and rupture are a concern
> **L**evofloxacin (Levaquin)

QUICK SUMMARY

Fluoroquinolone antibiotics inhibit DNA topoisomerase and DNA gyrase, preventing DNA coiling and causing DNA to break. Broad spectrum fluoroquinolones coverage includes Gram-positive, Gram-negative, and atypical bugs. You can give by PO or IV. Other QTc prolonging drugs carry an increased risk for QT prolongation and cardiac issues. Fluoroquinolones can cause *Clostridium difficile*. The **quinolone** stem is "oxacin, o-x-a-c-i-n" but **fluoroquinolones** have the "f-l" infix also.

One student of mine remembered that quinolones are for UTIs because Dr. Quinn, Medicine Woman, a 90s-television doctor on a show of the same name, is female. Women get proportionally more UTIs than men do.

Let's expand on our MEDICAL mnemonic.

M, moxifloxacin, has the best coverage against strep pneumonia – but not UTIs, as it does not reach adequate urine concentration. **Vigamox** is ophthalmic **moxifloxacin** with "v-i" for vision. Careful, Vigamox's 'amox, a-m-o-x' looks like amoxicillin, but it's not a penicillin. **Avelox** is moxifloxacin's oral or IV form.

E, extra careful with sunlight, as they have a phototoxicity risk.

D, delafloxacin meglumine is good for acute bacterial skin and skin structure infections (ABSSSI) such as cellulitis or wound infections. It's the first fluoroquinolone with MRSA and *Pseudomonas aeruginosa* coverage. It's non-inferior compared to vancomycin and aztreonam with more limited coverage than levofloxacin.

I, interact with cations, chelation with calcium, magnesium, aluminum, and iron. So, separate fluoroquinolones from antacids, multivitamins, or other medications or food containing these cations by two hours before or six hours after.

C, ciprofloxacin, has the greatest Gram-negative and pseudomonal activity. It's not a respiratory fluoroquinolone because it has no strep pneumonia activity. By cutting the "floxacin" stem from **ciprofloxacin**, the manufacturer made the brand name, **Cipro**.

A, Achilles tendonitis is a black box warning for Achilles rupture, especially in older patients and those also taking steroids.

L, levofloxacin, and ciprofloxacin have renal dose adjustments, but moxifloxacin does not. **Levofloxacin** is the left-handed (levo) isomer of another fluoroquinolone. The brand name combines the "lev" from **lev**ofloxacin and "quin" from fluoro**quin**olone to form **Levaquin**.

As was the case for fluoroquinolones, the next slide consists of medications covering Gram-negative bacteria.

IMMUNE MNEMONIC FLASHCARDS

Question 53. Identify the aminoglycoside Gram coverage, two toxicities, a method of assessment, it's mechanism of action, and five aminoglycoside drugs.

GRAM NEGATIVES

Gram-N_____ (Gram coverage)

R_____ Toxic

A_____ Toxic

M_____ P_____ (method of assessment)

N_____ / P_____ B
(aminoglycoside / polypeptide)

E

G_____ (aminoglycoside)

A_____ (aminoglycoside)

T_____ (aminoglycoside)

I_____ protein synthesis (30S) - mechanism of action

V

E

S_____ (aminoglycoside)

195

53. AMINOGLYCOSIDES – GRAM NEGATIVES

> **G**ram-negative
> **R**enal toxic
> **A**ura toxic (ototoxic to ears)
> **M**onitor peaks
>
> **N**eo<u>mycin</u> / Polymyxin B (Bacitracin)
> **E**
> **G**enta<u>micin</u> (Garamycin)
> **A**mi<u>kacin</u> (Amikin)
> **T**obra<u>mycin</u> (Tobrex)
> **I**nhibit protein synthesis (30S)
> **V**
> **E**
> **S**trepto<u>mycin</u>

QUICK SUMMARY

I think of the "side" (s-i-d-e) in aminogly<u>coside</u> and "cide, c-i-d-e" as in "cidal" to remind me that these drugs are bactericidal killers. The bacteria aminoglycosides cover form our mnemonic GRAM NEGATIVES.

G, Gram-negative, spectrum. Aminoglycosides interfere with protein synthesis and bind to the 30S ribosome portion. They cover pseudomonas or work in combination with beta lactams for synergy to fight staph and enterococci.

R, renal toxicity, do not give with other nephrotoxic drugs.

A, aura, a-u-r-a toxic, is ototoxicity. You might recognize the "A" for aura from Latin abbreviations like AU, both ears, AS, left ear, or AD, right ear.

M, monitor peaks. Aminoglycosides display concentration-dependent killing, so we dose them less often but at higher doses. We can dose aminoglycosides at normal intervals or in extended intervals, which use higher doses to achieve higher peaks. Therapeutic monitoring allows us to adjust dosages based on levels.

N, neomycin / polymyxin B, is a topical agent. Neomycin is the most toxic aminoglycoside to the kidney (nephrotoxic) and ears (ototoxic) when used systemically (throughout the body). However, patients can safely use *topical* preparations containing neomycin, such as over-the-counter **Neosporin**. The brand Neosporin takes "N-e-o and 's'" from neomycin sulfate, "p-o" from polymyxin B, and "r-i-n" from **Bacitracin.**

G, gentamicin, is an inexpensive choice. Just as practitioners shorten vancomycin as "vanc" in conversation, they condense gentamicin as "gent." The brand name **Garamycin** with "mycin, m-y-c-i-n" resembles gentamicin's "micin, m-i-c-i-n." The World Health Organization frowns on brand names with stem-like parts.

A, ami_kacin_, has the broadest spectrum and the least resistance.

T, tobra_mycin_, has an inhalable dosage formulation.

S, strepto_mycin_.

In the next slide, we specifically outline the aminoglycoside mechanism of action and side effects.

Question 54. Identify the aminoglycoside's Gram coverage, a method of assessment, its mechanism of action, and two toxicities.

AMINO

Activity (coverage) - A_____ G_____ N_____

M_____ P_____ (method of assessment)

I_____ protein synthesis (mechanism of action)

N_____ (side effect)

O_____ (side effect)

54. AMINOGLYCOSIDES MOA AND SIDE EFFECTS – AMINO

> **A**ctivity - aerobic Gram-negatives
>
> **M**onitor peaks
>
> **I**nhibit protein synthesis at the 30S ribosome
>
> **N**ephrotoxicity
>
> **O**totoxicity

QUICK SUMMARY

While I covered most of these in the GRAM NEGATIVES mnemonic, the AMINO mnemonic focuses less on drug names.

A, activity, with aerobic Gram-negative coverage.

M, monitor peaks for this concentration dependent antibiotic. Note, we also check troughs for toxicity.

I, inhibit protein synthesis at the 30S ribosome reminds us of the aminoglycoside's mechanism of action.

N, nephrotoxicity is damage to the kidney.

O, ototoxicity is damage to the ears.

From bactericidal drugs, we move to bacteriostatic medications.

Question 55. Name six bacteriostatic drugs or drug combinations.

STATIC

S_____ / T_____ (folic acid inhibitors)

T_____ (oxazolidinone)

A_____ (macrolide)

T_____ (tetracycline)

I_____ (nitrofuran)

C_____ (lincosamide)

55. BACTERIOSTATIC – STATIC

> **S**ulfamethoxazole / **T**rimethoprim (Bactrim)
>
> **T**edizolid (Sivextro)
>
> **A**zithromycin (Zithromax)
>
> **T**etracycline (Achromycin)
>
> N**i**trofurantoin (Macrobid, Macrodantin)
>
> **C**lindamycin – (Cleocin)

QUICK SUMMARY

Antibiotics are either **bactericidal** or **bacteriostatic**.

Bactericidal drugs kill bacteria and don't need the human immune system to help.

Bacteriostatic antibiotics are "bacterial birth control" preventing organisms from multiplying. They need the immune system to clear the bugs. Thus, we don't use bacteriostatics alone in immunosuppressed patients. Fittingly, the mnemonic is STATIC, S-T-A-T-I-C.

S, **sulfamethoxazole / trimethoprim**, a sulfa antibiotic.

T, **tedizolid**, represents the oxazolidinones.

A, **azithromycin**, represents the macrolides.

T, **tetracycline**, the drug class.

I, from **nitrofurantoin**, [[stress the "i" in ni"]] is a bacterial cell wall inhibitor only for uncomplicated UTIs because it concentrates below the kidney's level. It can cause rust-colored or brown urine discoloration. We dose **Macrobid** BID, twice daily and **Macrodantin** four times daily.

C, clindamycin, a lincosamide antibiotic, comes in many dosage forms and topical formulas. Systemic side effects include nausea and vomiting and a black-box warning of *Clostridium difficile*. It binds to the 50S ribosomal subunit to inhibit protein synthesis. Clindamycin covers aerobic and anaerobic bacteria, serves as a penicillin alternative in dental infections, and treats skin infections.

Most students remember the adverse effect CDAD (*Clostridium difficile*-Associated Disease) because there is a letter "c" and a "da, d-a" are after each other in the generic **clindamycin** name. Brand **Cleocin**, replaced the "i-n-d-a-m-y" in **clindamycin** with "e-o."

Next, we'll cover a sulfa antibiotic.

Question 56. With regards to sulfamethoxazole / trimethoprim, name a serious associated possible skin side effect condition, its primary therapeutic use, its spectrum of action, its mechanism of action, and an adjunct used to relieve UTI pain while it fights the infection.

SULFA

S_____ (skin condition)

U_____ (therapeutic use)

L_____ (spectrum of action)

F_____ A_____ - dihydrofolate reductase inhibitor

A_____ P_____ (adjunct for UTI pain)

56. Sulfamethoxazole / Trimethoprim - MOA and Side Effects - SULFA

> **S**tevens-Johnson Syndrome (SJS), a skin condition
>
> **U**TI – <u>u</u>rinary <u>t</u>ract <u>i</u>nfection medicine
>
> **L**ots of bacteria, broad spectrum
>
> **F**olic acid – dihydrofolate reductase inhibitor
>
> **A**ntiseptic phenazopyridine (Pyridium, Uristat)

Quick summary

Bactrim is made of two drugs – sulfamethoxazole and trimethoprim, abbreviated SMZ/TMP. Common side effects include rash and photosensitivity. It can form crystals in urine, so take with water. Watch for a warfarin interaction that can increase INR.

While sulfa drugs have "s-u-l-f-a" in them, some drugs have sulfa groups in the chemical's structure, but not in the generic name – for example, **furosemide.** I want to caution you about seeing sulfa moieties in chemical structures and assuming it will cause an allergic reaction. The academic literature doesn't support cross-sensitivity between allergies to sulfa antibiotics and other sulfonamide containing drugs like furosemide. Having a reaction to a loop with the history of antibiotic reaction means two distinct allergies. The brand name contains "b-a-c-t-r-i-m" from "**bacterium**" to make **Bactrim.**

Our mnemonic SULFA incorporates some of the side effects, spectrum of activity, and mechanism of action.

S, Stevens-Johnson Syndrome (SJS), a skin condition that is life threatening.

U, UTI – urinary tract infection medicine as its primary oral use.

L, lots of bacteria, broad spectrum including respiratory tract and MRSA skin infections. You can prophylax against *pneumocystis pneumonia* and toxoplasmosis in HIV patients.

F, folic acid – dihydrofolate reductase inhibitor. Sulfamethoxazole interferes with folic acid synthesis and trimethoprim inhibits the folic acid pathway. We base dosing on the trimethoprim component.

A, antiseptic phenazopyridine accompanies UTI treatments. It is not an antibiotic, rather an OTC urinary analgesic that turns urine red-orange.

From sulfonamides, we move to oxazolidinones.

Question 57. Name two resistant bacteria oxazolidinones show effectiveness against, a serious potential side-effect, two drug names, and the oxazolidinone class suffix.

MELT MRSA

M_____ and V_____ coverage

E_____ inhibitor (MAOI) to cause serotonin syndrome

L_____ (oxazolidinone)

T_____ (oxazolidinone)

Oxazolidinone class suffix: -z_____

57. OXAZOLIDINONES – MELT MRSA

> **M**RSA and VRE coverage
>
> **E**nzyme inhibitor (MAO) to cause serotonin syndrome
>
> **L**ine**zolid** (Zyvox)
>
> **T**edi**zolid** (Sivextro)

QUICK SUMMARY

The oxazolidinones, linezolid and tedizolid, bind the 50S subunit of the ribosome and inhibit protein synthesis. The "zolid, z-o-l-i-d" stem in **linezolid** comes from the **oxazolidinone** class. Side effects for oxazolidinones include nausea, diarrhea, a risk of optic and peripheral neuropathy, myelosuppression and decreased platelet count. They cover MRSA and VRE, therefore the MELT MRSA mnemonic.

M, MRSA and VRE coverage, an advantage of this drug class.

E, enzyme inhibitor of MAO to cause serotonin syndrome. Exercise caution when patients are also on antidepressants and similar serotonergic drugs.

Line_zolid_'s brand **Zyvox** ties to "solid" in "Man, Zyvox is zolid (solid). It treats two difficult-to-kill organisms – methicillin resistant staph aureus, MRSA, and vancomycin resistant enterococci, VRE."

Tedi_zolid_'s brand **Sivextro** sounds like it **vexes troublesome** MRSA and VRE.

Next, we tackle the macrolides.

Question 58. Name four macrolide antibiotics and their general antibiotic coverage.

FACE ATYPICAL BACTERIA

F_____

A_____

C_____

E_____

General antibiotic coverage: A_____

58. Macrolides – FACE ATYPICAL BACTERIA

> **F**idaxo<u>micin</u> (Dificid)
> **A**zi<u>thromycin</u> (Zithromax)
> **C**lari<u>thromycin</u> (Biaxin)
> **E**ry<u>thromycin</u> (EryPed)

Quick Summary

Macrolides interfere with protein synthesis and bind to the 50S ribosome portion. They provide good coverage of atypical bacteria like *legionella, chlamydia,* and *mycoplasma,* and we often use them for respiratory tract infections. Macrolides are time-dependent and bacteriostatic, inhibiting bacterial growth.

With hepatic elimination, patients don't need renal dosage adjustments. Caution patients about taking antacids with macrolides because they can cause chelation, making the antibiotics less effective. QT prolongation is possible, especially combined with other QTc prolonging drugs.

F, fidaxo<u>micin</u>, has a similar mechanism to macrolides, and it solely treats severe / resistant *Clostridium difficile*. The "dax" in the name **fi<u>dax</u>omicin** might come from its source, <u>Dactylosporangium</u> aurantiacum. The brand name **Dificid** suggests its primary therapeutic use against *Clostridium <u>difficile</u>*.

To recognize the three macrolides, **azithromycin, clarithromycin,** and **erythromycin,** you will see the "mycin, my-c-i-n" ending, but also a possible infix of "thro, t-h-r-o." Be careful: there *are* macrolides without this infix and stem like fidaxomicin.

A, azithromycin, has better Gram-negative coverage compared to erythromycin. The **Zithromax Z-pak** has two 250 mg tablets for a day one loading dose. Then patients take one tablet for the next four days. Or, we see one-time doses for sexually transmitted infections, STIs, or traveler's diarrhea, sometimes called the one-gram slam (1000 mg). Azithromycin has a different metabolic pathway than clarithromycin and erythromycin, avoiding the CYP enzyme interactions. The brand name **Zithromax** takes seven letters from **a**z**ithromyc**i**n** to build its name.

C, clarithromycin, has good Gram-positive coverage, and is often part of an *H. pylori* fighting multi-drug regimen. Again, you can use the "thromycin" to help clue you in to this drug's class. Gastroenterologists prescribe clarithromycin for peptic ulcer disease (PUD) triple therapy with amoxicillin and a proton pump inhibitor like omeprazole. The "b-i" in the brand name **Biaxin** indicates the twice daily dosing from the Latin abbreviation b.i.d. for *bis in die* – twice a day.

E, erythromycin, and clarithromycin have various drug-drug CYP enzyme metabolic interactions – including warfarin, an anticoagulant; theophylline, a methylxanthine for asthma; and carbamazepine, an antiepileptic. We rarely use erythromycin because of GI upset and multiple daily dosing. Instead, we use erythromycin to stimulate GI tract motility for gastroparesis.

Some **erythromycin** tablets are bright red like an erythrocyte is a red blood cell. Erythrocyte combines "erythro," Greek for "red," and "cyte," for cell. The brand **E-mycin** comes from taking the "rythro" out of the generic name **erythromycin** to render **E-mycin**.

We continue our look at drugs that affect Gram-positive organisms with the tetracyclines.

Question 59. Name three tetracycline antibiotics, a class suffix, and three interactions or side-effect concerns.

TIMED STAINED TEETH

T_____ (tetracycline)

I_____, D_____, M_____, A_____ (interactions)

M_____ (tetracycline)

E_____ S_____ (caution)

D_____ (tetracycline)

59. Tetracyclines – TIMED STAINED TEETH

> **T**etracycline (Achromycin)
> **I**ron, dairy, magnesium, aluminum chelation
> **M**inocycline (Minocin)
> **E**xcess sunlight or phototoxicity
> **D**oxycycline (Doryx)

Quick Summary

Tetracyclines have a similar mechanism of action to aminoglycosides and bind to the 30S ribosomal subunit, inhibiting protein synthesis. We mostly use tetracyclines for Gram-positive infections and respiratory flora. They also cover some unique pathogens. Tetracyclines may work for skin infections, acne, and community-acquired MRSA.

Pregnant or breastfeeding women and should avoid tetracyclines. Children under eight years old should not take tetracyclines because of the risk of bone growth suppression and tooth discoloration, so our mnemonic is TIMED STAINED TEETH.

T, tetracycline, has drug-drug interactions through CYP enzymes. Tetracycline, is essentially unavailable, but it's the class' namesake, so I included it.

I, iron, dairy, magnesium, and aluminum chelation with tetracycline antibiotics like minocycline and doxycycline.

M, mino<u>cycline</u>, like **doxycycline**, has the **tetracycline** class "cycline, c-y-c-l-i-n-e" stem. To create the brand name **Minocin**, the manufacturer dropped the "c-y-c-l" and last "e" from minocycline to get M-i-n-o-c-i-n, Minocin.

E, excess sunlight, or photosensitivity is a concern.

D, doxy<u>cycline</u>, has the "d" to remind you it treats periodontal disease in low doses. **Doryx**, the brand name, takes the first four letters of <u>doxy</u>cycline and adds an "r," scrambling them a bit to create Doryx.

From tetracylines, we go to tuberculosis.

Question 60. Name five medications for TB therapy.

RIPER

R_____

I_____

P_____

E_____

R_____

60. Antituberculosis Agents – RIPER

> **R**ifampin (Rifadin)
>
> **I**soniazid (INH)
>
> **P**yrazinamide (PZA)
>
> **E**thambutol (Myambutol)
>
> **R**ifabutin (Mycobutin)

Quick summary

Mycobacterium tuberculosis causes tuberculosis, or TB. It's a contagious respiratory infection spread by water droplets. We use the RIPER mnemonic, referring to a positive TB test's raised induration which looks ripe.

R, rifampin, also known as rifampicin, can cause benign flu-like symptoms and red-orange saliva, sweat, and tears. Rifampin interacts with warfarin, protease inhibitors for HIV, and oral contraceptives. If HIV patients use a protease inhibitor, they should substitute the 2nd 'R' in the RIPER acronym, rifabutin. Students remember that rifampin turns secretions like tears, sweat, and urine red with its first letter "r." The brand name **Rifadin** replaces the "m-p" from rifampin with a "d" to make Rifadin.

I, isoniazid, is for active and latent TB. Watch for upset stomach, increased LFTs, and neuropathy – especially in diabetics, HIV patients, and alcoholics. Pyridoxine, vitamin B6, reduces this risk. The "n" in **isoniazid** reminds students of neuropathy, an adverse effect. There is no "H" in **isoniazid**, so the acronym **I-N-H** comes the chemical name, **i**so**n**icotinyl **h**ydrazide. INH.

P, pyrazinamide, which starts with "p" reminds us adverse effects include polyarthritis and gout, another kind of arthritis. The **P-Z-A** abbreviation pulls three letters from generic p̲yra̲zina̲mide.

E, ethambutol, can cause optic neuritis, leading to decreased vision and color blindness. The "e" for "eyes" or "o" for optics in e̲thambuto̲l helps us remember. The m-y begins brand **Myambutol** because M̲y̲cobacterium tuberculosis is the causative agent.

R, rifabutin, is a substitute for rifampin in patients taking protease inhibitors.

Next, we move on to hepatotoxic drugs.

Question 61. Name four drugs that can cause hepatotoxicity.

LIVER PAIN

P_____ (antituberculosis)

A_____ (non-narcotic analgesic)

I_____ (antituberculosis)

N_____ (analgesic/antiinflammatory drug class)

61. Hepatotoxicity – LIVER PAIN

> **P**yrazinamide (PZA)
>
> **A**cetaminophen (Tylenol)
>
> **I**soniazid (INH)
>
> **N**SAIDS

Quick Summary

These are just a few hepatotoxic medications. Although you don't always feel physical pain with mild hepatotoxicity, your liver feels it. Monitor LFTs and signs and symptoms. Note, NSAID hepatotoxicity is specific to the individual drug. From antibacterials, we move to antifungals.

Question 62. Name thirteen antifungals and the molecule related to cholesterol that many antifungals attack.

ANTI-FUNGAL EVICTIONS

A_____ (polyene antifungal)

N_____ (polyene antifungal)

T_____ (allylamine)

I_____ (triazole)

F_____ (triazole)

FlU_____ (antimetabolite pyrimidine analog)

FuN_____ (echinocandin)

G_____ (miscellaneous antifungal)

PosA_____ (triazole)

L_____ (imidazole)

E_____ (molecule related to cholesterol that many antifungals attack)

V_____ (triazole)

MI_____ (imidazole)

C_____ (imidazole)

T_____ (triazole)

62. ANTIFUNGALS – ANTI-FUNGAL EVICTIONS

Amphotericin B (Fungizone)

Nystatin (Mycostatin, Nystop)

Terbinafine (Lamisil)

Itraconazole (Sporonox)

Fluconazole (Diflucan)

Fl**U**cytosine (Ancobon)

Fu**N**gins (echinocandins)

Griseofulvin (Grifulvin, Gris-PEG)

Pos**A**conazole (Noxafil)

Luliconazole (Luzu)

Ergosterol

Voriconazole (Vfend)

M**i**conazole (Monistat-3)

Clotrimazole (Mycelex)

Terconazole (Terazol 7) *(EVICTIONS)*

QUICK SUMMARY

You can get a little crazy with the mnemonics, and with anti-fungal evicts, we tie fourteen antifungals together in a single phrase. You can picture the antifungal landlord giving a fungus the boot with the mnemonic ANTI-FUNGAL EVICTIONS.

Mnemonic Part 1. ANTI

A, amphotericin B, binds ergosterol, our 'E' in this mnemonic, a cholesterol-like product in fungi cell membranes. That binding alters cell membrane permeability to cause cell death. There are two amphotericin formulations, conventional and multiple liposomal. The conventional form, amphotericin B deoxycholate, has many side effects, including infusion reactions and nephrotoxicity and requires premedication with acetaminophen and diphenhydramine. Students ask, what happened to **amphotericin A**? It was ineffective, so scientists invented **amphotericin B**. The brand **Fungizone**, makes it easy to remember it's an antifungal.

N, nystatin, also binds ergosterol, poking holes in cell membranes and leading to fungal death. It's available as a suspension, tablet, and topical. We often use it for mild thrush caused by *Candida albicans*. Patients should swish and spit, and not swallow if possible, as nystatin can cause nausea and stomach pain. Esophageal thrush may need swallowing. Nystatin ends in "statin." A class of cholesterol lowering drugs, the HMG-CoA reductase inhibitors, are "statins" because they end in s-t-a-t-i-n. A better substem + suffix stem for HMG-CoA reductase inhibitors is "vastatin, v-a-s-t-a-t-i-n." To keep from thinking **nystatin** was ever a cholesterol lowering "statin," a student remembered the dosage forms nystatin comes in: powder and liquid to swish and spit.

T, terbinafine, inhibits squalene epoxidase, a sterol biosynthesis enzyme in fungi. This results in deficient ergosterol in the fungal cell membrane, causing cell death.

I, itraconazole, is an azole antifungal that decreases ergosterol synthesis covering different fungal infections. All azole antifungals can increase liver function tests, LFTs.

Itraconazole treats various *Candida* species and aspergillus. It has pH-dependent absorption. An increase in pH causes a decrease in absorption, so separate antacids and PPIs from doses. The capsules and oral solution are not interchangeable. We take the solution without food, and the capsules and tablets with food. Watch for other medications that cause QTc prolongation.

Mnemonic Part 2. FUNGAL.

F, fluconazole, has a narrow spectrum of action. It covers *Candida albicans* well and we see it as a single dose for vaginal candidiasis. It covers other candida species poorly. Fluconazole needs dosage adjustments in renal failure. The "conazole, c-o-n-a-z-o-l-e" ending helps identify fluconazole as an antifungal drug. What the "f-l-u" in fluconazole for fluorine atom, not influenza. One student used the first three letters of the brand **Diflucan** as the forceful command, "**Die fungi!**"

U, FlUcytosine.

N, fungin, echinocandins are for invasive fungal infections, especially for candida infections. Common ones are micafungin and caspofungin

G, griseofulvin: take with a fatty meal.

A, PosAconazole, is active against aspergillus and zygomycetes molds. The suspension and tablet are not interchangeable. Dosing regimens differ as tablets have better absorption. Side effects include hypokalemia, hypomagnesemia, and nausea.

L, Luliconazole.

Antifungal Part 3. EVICTIONS

E, ergosterol, a cholesterol like compound, is often the target of antifungals and a vehicle for side effects because it resembles cholesterol.

V, voriconazole, is the drug of choice in aspergillus, but can cause liver damage and visual changes like blurred vision and altered color perception. Don't refrigerate the suspension, and shake well before use. There are many drug-drug interactions with voriconazole, and it exhibits linear and then non-linear kinetics, so small dose changes can lead to major level changes.

I, mIconazole.

C, clotrimazole, inhibits ergosterol synthesis. Like nystatin, we use it for mild *candida* oropharyngeal infections. It's available as a lozenge and troche in topical and vaginal forms. Clotrimazole can cause nausea.

T, terconazole.

It's not in the mnemonic, but we rarely use **ketoconazole** (**Nizoral**) anymore. It has many CYP drug-drug interactions and causes hepatotoxicity.

From the antifungals, we move to antivirals.

Question 63. Name three drugs that affect the herpes simplex virus.

SAVED PROM

S

A_____ (prescription)

V_____ (prescription)

E

D_____ (over-the-counter)

63. ANTIVIRALS – HERPES SIMPLEX VIRUS & VARICELLA ZOSTER VIRUS (HSV/VZV) – SAVED PROM

> S
>
> **A**cyclovir (Zovirax)
>
> **V**alacyclovir (Valtrex)
>
> E
>
> **D**ocosanol (Abreva)

QUICK SUMMARY

Drugs for herpes infections such as **acyclovir** and **valacyclovir** can help prevent recurrences and treat an infection, but they do not cure the disease. **Docosanol** is an over-the-counter topical for cold sores that completes our SAVED PROM mnemonic.

S,

A, acyclovir treats Varicella-Zoster virus (VZV) and herpes simplex virus (HSV). You can think of brand **Zovirax** as a drug that axes Zoster virus. Dosing is five times daily.

V, valacyclovir has acyclovir in the root because it's the valine ester. Valacyclovir, a prodrug, turns into active acyclovir in the body. Valacyclovir allows for twice daily dosing for patient compliance. The brand name, **Valtrex,** includes the "val" from valacyclovir plus T-rex, and wrecks a virus.

E,

D, docosanol speeds recovery from bothersome cold sores. Finding this product often presents a challenge. It's sometimes highlighted at the end of the aisle on leftover shelf space. The blue tube is smaller than your pinky and comes in a hard-plastic clear package.

I wondered, "Who would spend roughly $20 on such a small tube?" But then I thought, "Well if I was going to prom and it's the only time I ever get to attend, then I would probably pay the money."

With this idea, I thought of how **docosanol** lets you go to the ball, and this drug class SAVED PROM. The brand name **Abreva** is a little easier to remember. It helps abbreviate or shorten the impact of a cold sore.

Next, we cover influenza virus medications.

Question 64. Name three medications patients can use for influenza after acquiring it, a side-effect of an influenza medication and an unusual dosage form for one of the drugs.

FLU ZOMBIE

Z_____ (antiviral)

O_____ (antiviral)

M_____ / D_____ (side effect for oseltamivir)

B

I_____ (dosage form for zanamivir)

pE_____ (antiviral)

64. ANTIVIRALS – INFLUENZA – FLU ZOMBIE

> **Z**anamivir (Relenza)
> **O**selt**a**mivir (Tamiflu)
> **M**ood / Delirium like a zombie *(Oseltamivir)*
> **B**
> **I**nhaled *(Zanamivir)*
> p**E**ramivir injection (Rapivab)

QUICK SUMMARY

Many antivirals, including these influenza drugs, have a "-vir-" stem. Our mnemonic FLU ZOMBIE is how you feel when you have the flu.

Z, zanamivir comes in a Diskhaler, a way to get powder to the lungs. The Diskhaler is difficult for patients with dexterity issues, but provides an alternative to **oseltamivir**. With zanamivir's brand name think: **Relenza** "represses influenza" virus or Relenza makes "influenza relent" (give up).

O, oseltamivir may be a medication all family members get if one person is sick enough or a family member is immunocompromised. The brand **Tamiflu** alludes to a drug that "tames the flu." Treatment dosing is twice daily for five days for active influenza and once daily for ten days for prophylaxis. Both zanamivir and oseltamivir work when taken within 48 hours of the infection.

M, mood. Both zanamivir and oseltamivir can cause abnormal behavior or unusual mood changes.

B,

I, inhaled, is zanamivir's dosage form.

E, pEramivir injection is single-dose emergency department (ED) treatment for those for whom oral medicines are not suitable. Peramavir's brand name **Rapivab** has the first four letters of "rapid" and its dosage form, "IV" in the middle.

From influenza virus, we move to HIV treatments.

Memorizing Pharmacology Mnemonics

Question 65. Name seven HIV medications in order of attack to the host cell.

THEM THREE, RAID HIV

T_____ and quad therapy

HIV

E_____ (Fusion inhibitor)

M_____ (CCR5 receptor antagonist)

T_____ (NRTI)

H

R

E_____ (NNRTI)

E_____ (NNRTI)

Ra_____ (Integrase inhibitor)

D_____ (Protease inhibitor)

65. HIV ANTIVIRALS – THEM THREE, RAID HIV

> Triple and quad therapy
>
> **HIV**
>
> **E**nfu<u>virtide</u> (Fuzeon) Fusion inhibitor
>
> **M**ara<u>viroc</u> (Selzentry) CCR5 receptor antagonist
>
> **(All three as ATRIPLA)**
>
> **T**enofo<u>vir</u> (Viread) NRTI (reverse transcriptase)
>
> **H**
>
> **R**
>
> **E**fa<u>virenz</u> (Sustiva) NNRTI (reverse transcriptase)
>
> **E**mtri<u>citabine</u> (Emtriva) NNRTI (reverse transcriptase)
>
> **Ra**<u>ltegravir</u> (Isentress) Integrase inhibitor
>
> **D**aru<u>navir</u> (Prezista) Protease inhibitor
>
> *In order of attack on the infected cell*

QUICK SUMMARY

HIV drugs affect specific targets in the cell or retrovirus. HIV medications, like tuberculosis medications, work best in drug combinations. I've organized the five HIV drug classes in the order an HIV virus attacks a healthy cell. First, the HIV virus tries to fuse with the cell, and then it uses cellular chemokine receptor five (CCR5) to enter the cell. Inside the cell, the HIV virus uses reverse transcriptase, integrase, and protease. HIV medications have three-letter abbreviations, as these drugs are not only hard to pronounce, but conversation filled with several multisyllable words can make understanding difficult. The mnemonic TIMES THREE, RAID HIV reminds us that we use multiple medications to avoid resistance.

T, triple and quad therapy reminds us we use multiple drugs at a time.

H, HIV is the target virus.

E, enfuvirtide's generic contains "vir" for antiviral. Brand **Fuzeon** is a fusion inhibitor.

M, miraviroc's "-vir-" stem suggests antiviral, but the sub-stem "-viroc" has a "c" at the end for CCR5 antagonist. Think of **maraviroc** as a "rock" guarding against viral entry. The brand name **Selzentry** sounds a lot like a "sentry," a guard.

T, tenofovir,

H,

R,

E, efavirenz, and

E, emtricitabine, form **Atripla**. To remember complex names, try to memorize the stems "-vir, "-virenz," "-citabine," first. Then add the other two or three syllables to memorize the whole names of **tenofovir, efavirenz,** and **emtricitabine**. Slow down here, it takes time to work with unfamiliar names. The brand name **Atripla** refers to the three drugs, a "triple" surrounded by two A's that can stand for "against AIDS."

RA, raltegravir is an integrase strand transfer inhibitor. Inside the generic name, you find the stem "-tegravir" made up of "tegra," a part of integrase, and "vir" for antiviral. The brand name **Isentress** also looks like "I" for integrase and "s-e-n-t-r" from "sentry."

I,

D, darunavir's Prezista sounds like resist, spelled "r-e-z-i-s-t," with the first two letters "p-r" of "protease." Darunavir's a protease inhibitor.

From adults and potentially children with HIV we move to RSV in pediatrics.

Question 66. Name a medication for RSV.

Medication for treating Respiratory Syncytial Virus (RSV)

_____

66. ANTIVIRAL – RESPIRATORY SYNCYTIAL VIRUS – PEDIATRIC PALIVIZUMAB FOR PREVENTION

Palivizu<u>mab</u> (Synagis)

QUICK SUMMARY

Respiratory syncytial virus (RSV) is usually unproblematic in healthy adults, but in infants younger than one year old, it can be deadly. A drug like **palivizumab** can prevent RSV in at-risk patient populations.

The prefix "p-a-l-i" has "P" and "I" for pediatrics or infants at risk for RSV. In palivizumab, the

"pali" is a prefix that separates it from similar drugs.

"-vi-" stands for antiviral (the target),

"-zu-" stands for humanized (the source), and

"-mab" is for <u>m</u>onoclonal <u>a</u>nti<u>b</u>ody.

This biologic stem + infixes resemble **infliximab (Remicade)** for ulcerative colitis or **omalizumab (Xolair)** for asthma, but with a different clinical purpose. The brand **Synagis** looks like "synergy" which rhymes with RSV.

From RSV, we move to Hepatitis-C medication classes.

Memorizing Pharmacology Mnemonics

Question 67. Name five medications for hepatitis-C.

DAILY DOSES

D_____ (NS5A inhibitor)

O_____ (NS5A inhibitor)

S_____ (Protease inhibitor)

E_____ (NS5A inhibitor)

S_____ (NS5B inhibitor)

67. THREE HEP-C VIRUS (HCV) MEDS – DAILY DOSES

> **D**aclat<u>asvir</u> (Daklinza)
>
> **O**mbit<u>asvir</u> (in combination with other drugs)
>
> **S**ime<u>previr</u> (Olysio)
>
> **E**lb<u>asvir</u> (Zepatier)
>
> **S**ofos<u>buvir</u> (Sovaldi)

QUICK SUMMARY

While hepatitis-C regimens show exceptional success, their complicated regimens are beyond this book's scope. With most HCV regimens, our mnemonic, DAILY DOSES applies.

D, daclat<u>asvir</u>,

O, ombit<u>asvir</u>, and **elb<u>asvir</u>** inhibit the NS5A HCV protein. The -asvir, a-s-v-i-r stem signals nonstructural protein 5A (NS5A) inhibitors.

S, sime<u>previr</u> is an NS3/4A serine protease inhibitor stopping protein synthesis based on its previr, p-r-e-v-i-r stem.

E, elb<u>asvir</u> another NS5A HCV protein inhibitor fits in the mnemonic here.

Sofos<u>buvir</u>, is an RNA polymerase (NS5B) inhibitor based on its buvir, b-u-v-i-r stem.

From the immune section, we move on to neuro / psych.

CHAPTER 5: NEURO/PSYCH MNEMONIC FLASHCARDS

Question 68. Name ten sedative-hypnotic drugs or drug combinations, identifying the over-the-counter option, three benzodiazepine-like drugs, a melatonin receptor agonist, three benzodiazepines, an orexin receptor antagonist, and an antidepressant.

A PEZZ PILLOW RESTTT

A_____/ D_____ (OTC)

P

E_____ (non-benzodiazepine)
Z_____ (non-benzodiazepine)
Z_____ (non-benzodiazepine)

R_____ (melatonin receptor agonist)
E_____ (benzodiazepine)
S_____ (orexin receptor antagonist)
Te_____ (benzodiazepine)
Tr_____ (benzodiazepine)
Tr_____ (antidepressant)

68. Sedative/Hypnotics – A PEZZ PILLOW RESTTT

Acetaminophen / Diphenhydramine (Tylenol PM)

P

Eszopiclone (Lunesta)

Zaleplon (Sonata)

Zolpidem (Ambien, Ambien CR)

Ramelteon (Rozerem)

Estazolam (ProSom)

Suvorexant (Belsomra)

Temazepam (Restoril)

Triazolam (Halcion)

Trazodone (Soma)

Quick Summary

Let's start with a couple of questions.

What separates a sedative from a hypnotic? Often the dose. A sedative calms someone down, a hypnotic helps them sleep.

How do meds help insomniac patients? Some agents help them fall asleep, others help them stay asleep, and others do both.

We'll use a candy that looks like a pillow for our mnemonic: A PEZZ, P-E-Z-Z, PILLOW RESTTT, R-E-S-T-T-T.

Mnemonic Part 1. A PEZZ.

A, acetaminophen / diphenhydramine, combines a non-opioid analgesic plus a sedating first-generation antihistamine. One student said "take both 'phens,' acetamino*phen* and di*phen*hydramine, to end up sleepin'".

Which patient population(s) should avoid antihistamines like diphenhydramine as sleep aids? The elderly based on the Beers Criteria because of potential confusion & hangover effects. Also, children because these drugs might *cause* stimulation and insomnia. From Chapter 3, diphenhydramine is an anticholinergic, with dry mouth, constipation, blurred vision, and confusion as side effects in the elderly.

What, then, do you recommend for the elderly and children? Good sleep hygiene and melatonin.

PEZZ, with two Z's, represents **the** non-benzodiazepines eszopiclone, zaleplon, and zolpidem which are first-line over benzodiazepines. Why? Even though non-benzodiazepines are C-IV substances, they have less dependence risk and 'hangover' effect which can happen if patients don't get the recommended 7 to 8 hours of sleep. Other side effects include dizziness, drowsiness, and, rarely, unusual actions while sleeping like sleepwalking and sleep driving.

E, eszopiclone, is for sleep onset or maintenance. Eszopiclone's generic stem '-clone, c-l-o-n-e' "will put you in the sleeping zone" or look to the "z" in eszopiclone for getting your z's. Brand **Lunesta** uses the Latin for the moon, "Luna," plus "est, e-s-t" from "rest." Lunesta.

Z, zaleplon, is short-acting with a fast onset, so it's best for patients having trouble falling asleep. Zaleplon's brand **Sonata** comes from the Italian sonare [[Italian accent]], to sound, as in "Sonata is music to sleep by."

Z, zolpidem, has an immediate release form for sleep onset and extended-release for onset and maintenance. Novel dosage forms include sprays and sublingual tablets. Use the "-pidem, p-i-d-e-m" stem to remember **zolpidem** as a sedative-hypnotic and the "z" in **zolpidem** – for getting z's just like with eszopiclone. Match brand **Ambien** with an ambient, sleepy environment. Look for **zolpidem** controlled release, brand **Ambien C-R**, for patients with difficulty maintaining sleep (D-M-S) *and* difficulty falling asleep (D-F-A). Regular zolpidem only works for those with difficulty falling asleep, D-F-A.

Mnemonic Part 2. RESTTT (with three Ts)

R, ramelteon, a melatonin receptor agonist for sleep onset insomnia is not a controlled substance. Avoid fatty food as this can alter absorption. It can cause vivid dreams. Since melatonin helps regulate circadian rhythms, for whom might you recommend it? Jet-lagged travelers and overnight shift workers. The "–melteon, m-e-l-t-e-o-n" stem in **ramelteon** lets you know it's a melatonin agonist. A student of mine said the "m-e-l" in **ramelteon** reminded her of the word mellow, m-e-l-l-o-w. In

Rozerem, you can see the "z" for z's, the "r-e-m" for rapid eye movement REM sleep, and "roze" rhymes with doze.

E, est<u>azolam</u>, has the –azolam benzodiazepine stem and its brand **ProSom** combines "provide" and "somnolence." Somnolence means sleepiness.

S, suv<u>orexant</u>, is a dual orexin receptor antagonist for sleep onset or maintenance. Suvorexant has the -orexant stem signaling its chemical class of orexin antagonists. Its brand **Belsomra** combines "bella" and "somnolence" which translates to "beautiful sleep."

T, tem<u>azepam</u>, a benzodiazepine, has a specific insomnia indication. How does temazepam's metabolism make it safer? The body clears it hepatically, so it's a safer choice for the elderly and patients with renal impairment. We'll talk about this in the LOT, L-O-T mnemonic of lorazepam, oxazepam, and temazepam later on. Temazepam's brand **Restoril** uses "<u>rest</u>" or "<u>restore</u>" as its brand name.

T, tri<u>azolam</u>, can cause anterograde amnesia. What does anterograde mean? The root 'ante' means 'before.' If you're a poker player, the ante comes out before the cards; your forearm is your antebrachium, that which comes 'before' your arm. Anterograde amnesia is the inability to form new memories – you lose memories 'before' they even form. While brand **Halcion**, has an "i" in it, the word halcyon, h-a-l-c-y-o-n means a past peaceful time. Many parents look to a time before kids as halcyon.

T, tr<u>azodone</u>, is an antidepressant that inhibits serotonin reuptake and histamine and alpha receptors. Insomnia guidelines don't currently recommend its off-label use. Sexual dysfunction and orthostasis are side effects. What patient might value trazodone? A depressed *and* insomniac patient benefits from the antidepressant and sedative-hypnotic combination. In practice, **trazodone** helps patients sleep and many students look to the "z" in trazodone to remember its role.

Now, let's look at drugs which can reduce anxiety.

Question 69. Name a non-benzodiazepine anxiolytic, a concern with its speed of action, and four benzodiazepine anxiolytics.

BE CALM

B_____ (non-benzodiazepine)

E_____ (concern)

C_____ (benzodiazepine)

A_____ (benzodiazepine)

L_____ (benzodiazepine)

M_____ (benzodiazepine)

69. ANXIOLYTICS - BE CALM WITH BUSPIRONE AND BENZOS

> **B**uspirone (Buspar) is not a benzo!
> **E**xtended time to start working
>
> **C**lonazepam (Klonopin)
> **A**lprazolam (Xanax)
> **L**orazepam (Ativan)
> **M**idazolam (Versed)

QUICK SUMMARY

Our BE CALM mnemonic reminds us that these drugs work to reduce anxiety.

Mnemonic Part 1. BE

B, buspirone, is a non-benzodiazepine with no effect on GABA, and an unclear mechanism of action. We think it has an affinity for serotonin receptors. As nonfiction literature is "not" fiction, we classify buspirone as a non-barbiturate, non-benzodiazepine. Side effects include nausea, dizziness, headache, and drowsiness. Patients take without regard to food and consistently from day to day. Watch for serotonin syndrome with other serotonergic agents.

E, extended time to start working reminds us buspirone takes weeks to months to become effective. PRN benzodiazepines can help control acute attacks.

Mnemonic Part 2. CALM

Benzodiazepines are controlled substances that potentiate GABA to cause central nervous system depression. This helps with anxiety, convulsions from seizures, and sleep. However, benzodiazepines often fail the social anxiety patient. Benzos cause drowsiness to dampen thinking. A better choice is a beta-blocker that reduces tremors, tachycardia, and palpitations in performance. Let's look now to four benzodiazepines.

C, clonazepam, has the "azepam, a-z-e-p-a-m" benzodiazepine stem. Some online drug cards say the ending is just "p-a-m, pam." This is wrong. **Fenoldopam** ends with –pam, but is a D2 agonist, not a benzodiazepine. Brand **Klonopin** uses the phonetic spelling of clonazepam's first four letters, "c-l-o-n."

A, alprazolam, has an '-azolam' stem signaling benzodiazepine also. Alpr*az*olam has one z; benzodiazepine has two; brand **Xanax** sounds like a "z" to help you get a snooze and "x's" out an*x*iety too, is one student mnemonic.

L, lorazepam, we see for alcohol withdrawal syndrome. It clears hepatically and is a good option for elderly and renal failure patients. Remember lorazepam through the "–azepam" stem or think about brand **Ati-*van*** *van*quishing anxiety.

M, midazolam has two M's for the <u>m</u>emories you can't form since it causes anterograde amnesia. Again, your *ante*brachium is your forearm, and the *ante* is the money poker players put out *before* the dealer deals. *Ante*rograde amnesia is the inability to form future memories. Alternatively, you can use "I can't remember the 'verse you just said'" for brand **Versed**.

Next, we'll review specific benzodiazepine properties.

Memorizing Pharmacology Mnemonics

Question 70. Identify seven benzodiazepines as short, intermediate, or long-acting.

Short

M_____

Tr_____

Intermediate

A_____

L_____

Te_____

Long

Ch_____

Cl_____

D_____

70. Benzodiazepines – short to long-acting

> **S**hort
> **M**idazolam (Versed)
> **T**riazolam (Halcion)
>
> **I**ntermediate
> **A**lprazolam (Xanax)
> **L**orazepam (Ativan)
> **T**emazepam (Restoril)
>
> **L**ong
> **C**hlordiazepoxide
> **C**lonazepam (Klonopin)
> **D**iazepam (Valium)

Quick summary

It's important to know how long benzodiazepines last in the body. For example, by knowing which drug has a long half-life, like chlordiazepoxide, we can predict its use in alcohol withdrawal. When there's not an obvious mnemonic, sort them alphabetically as I've done. Besides knowing relative benzodiazepine half-lives, we should also know which undergo hepatic metabolism which we'll cover next.

Memorizing Pharmacology Mnemonics

Question 71. Identify three benzodiazepines with extra-hepatic metabolism.

OUTSIDE LOT

Benzodiazepines with extra-hepatic metabolism

L_____

O_____

T_____

71. Extrahepatic benzodiazepine metabolism— OUTSIDE LOT

> Lorazepam (Ativan)
>
> Oxazepam (Serax)
>
> Temazepam (Restoril)

Quick Summary

We may need a benzodiazepine that undergoes phase II metabolism. When the body metabolizes the drug outside of the liver, it's a safer choice for hepatic disease and we prefer them in the elderly.

Let's move on to antidotes including one for benzodiazepines.

Question 72. Name eight antidotes and their target agent(s).

FIND AND PACK ANTIDOTES

Antidote > Target agent

F_____ for B_____

I_____ for D_____

N_____ for O_____

D_____ for D_____

P_____ for H_____

A_____ for A_____

C_____ for A_____

K_____ for W_____

72. Eight Antidotes – FIND AND PACK ANTIDOTES

> *F*lumazenil (Romazicon) for benzodiazepines
>
> *I*darucizumab (Praxbind) for dabigatran (Pradaxa)
>
> *N*aloxone (Narcan) for opioids
>
> *D*igoxin Immune Fab (Digibind) for digoxin
>
> *P*rotamine sulfate for heparin
>
> *A*cetylcysteine (Mucomyst) for acetaminophen
>
> *C*holinesterase inhibitor (physostigmine) for anticholinergics
>
> *K* vitamin phytonadione (Mephyton) for warfarin

Quick Summary

An overdose needs quick knowledge of the right antidote or reversal agent. While there are others, I want to point out a pipeline drug, andexanet alfa, which might be FDA approved soon. It reverses Xa inhibitors. Scientists have studied it in rivaroxaban and apixaban.

Next, we'll talk about a medication class that has an overdose concern - barbiturates.

Question 73. Name three barbiturates, a class suffix, and then which organ's function they can suppress.

BARB SLEPT TOO LONG

S_____ (barbiturate)

L_____ (organ)

E

P_____ (barbiturate)

T_____ (barbiturate)

Class infix

-b_____

73. Barbiturates – BARB SLEPT TOO LONG

> **S**eco<u>barb</u>ital (Seconal)
>
> **L**ungs, watch for respiratory depression
>
> **E**
>
> **P**heno<u>barb</u>ital (Luminal)
>
> **T**hiopental (Pentothal)

Quick summary

The mnemonic BARB SLEPT TOO LONG helps us remember the drug stem barb, b-a-r-b, and that barbiturates are dangerous potentially causing deadly respiratory depression.

S, seco<u>barb</u>ital, is a DEA Schedule II agent anesthetic and anticonvulsant. Veterinarians use it to euthanize pets.

L, lungs, watch for respiratory depression

E,

P, pheno<u>barb</u>ital, is a DEA Schedule IV medication with oral, IM, and IV dosage forms. I put the barbiturate pheno<u>barb</u>ital after benzodiazepines because we don't use them much anymore.

T, thiopental, is rapid-onset, a short-acting general anesthetic.

Let's make sure we know the differences between benzodiazepines and barbiturates with a few quick questions.

1. **Why do prescribers prefer benzodiazepines to barbiturates?** Barbiturates can cause respiratory depression and residual sedation, sleeping too long, and they have a narrow therapeutic window. Tolerance can develop to chronic barbiturate use. The present clinical practice employs non-benzodiazepines and benzodiazepines for insomnia. We see prescribers use barbiturates sparingly for seizure disorders and anesthesia.
2. **How does tolerance make barbiturates' narrow therapeutic window more concerning?** Tolerance leads to increased doses and toxicity or death. Barbiturates can independently open GABA channels whereas benzodiazepines prolong the opening once naturally opened.
3. **What side effects do we need to watch for?** Side-effects include sedation, dizziness, nausea, and vomiting. There is also the potential for Stevens-Johnson Syndrome, a rare and sometimes fatal skin rash.

Next, we'll address drugs related to quitting smoking.

Question 74. Name two medications for smoking cessation and a single side-effect or disease-drug interaction for both.

STOP SMOKING VIBE

V_____ (medication)

I _____ (side effect)

B_____ (medication)

E_____ (drug-disease interaction)

74. Smoking Cessation – STOP SMOKING VIBE

> **V**arenicline (Chantix)
> **I**deations and dreams
> **B**upropion (Zyban)
> **E**pileptics beware

Quick Summary

It often takes many tries to quit smoking, but these two drugs help patients get the STOP SMOKING VIBE.

V, varenicline, is a nicotine receptor agonist causing low-level nicotine receptor stimulation. This action prevents nicotine binding and prevents withdrawal symptoms. How does a patient use varenicline? They start a week before a quit date, then taper out over the next 12 weeks. The patient can repeat to ensure they quit.

Varenicline has the "nicline, n-i-c-l-i-n-e" stem. And if you replace that "l" with an o-t, you get n-i-c-o-t-i-n-e, nicotine. A student, in a southern drawl, remembered varenicline's generic name by saying, "With **varenicline**, 'I'm vary incline ta quit.'" Another came up with "My chant is, 'I don't need my fix' with **Chantix**."

Can we expect neuropsychiatric side-effects with varenicline?

I, ideations and dreams, were reported early on, but the manufacturer removed the boxed warning in December of 2016.

B, bupropion, is a twice-daily smoking cessation drug initially marketed as an antidepressant. Bupropion blocks dopamine and norepinephrine reuptake, reducing cravings. We use only the sustained-release formula for smoking cessation, but the immediate and sustained release form for depression.

We start bupropion a week before the quit date for up to six months. Advantages include a lack of weight gain and less risk of sexual dysfunction. Unlike varenicline, patients can take bupropion with nicotine replacement gum, patches, or lozenges.

The manufacturer first marketed bupropion as **Wellbutrin**, combining "well" from wellness and "bu, b-u" from bupropion as an atypical antidepressant. This antidepressant group doesn't fit into the SSRIs, SNRIs, TCAs, or MAOIs. Depressed patients that smoked must have quit, so the manufacturer rebranded bupropion as **Zyban**, to put a "ban" on smoking.

What concerns us most about bupropion?

E, epileptics, must beware of bupropion reducing seizure threshold.

Now, let's look at drugs for ADHD.

Question 75. Name two non-stimulant ADHD agents, four single or combination stimulant ADHD drugs, and two types of formulations

CALMED ADHD

C_____ (non-stimulant)

A_____ (non-stimulant)

L_____ (stimulant)

M_____ (stimulant)

E

D_____ (stimulant)

A _____ /

D_____ (stimulant combination)

H

D_____ formulations, _____

and _____.

75. ADHD Agents – CALMED ADHD

> **C**lonidine (Catapres, Kapvay)
>
> **A**tom**oxetine** (Strattera)
>
> **L**isdexam**fetamine** (Vyvanse)
>
> **M**ethylphenidate (Ritalin, Concerta, Metadate, Quillivant)
> **E**
> **D**exmethylphenidate (Focalin)
>
> **A**m**phetamine**/
> **D**extroam**phetamine** (Adderall)
> **H**
> **D**ifferent formulations, short and long-acting

Quick Summary

Attention-deficit, hyperactivity disorder, ADHD, happens when a child or adult has continued inattention and/or hyperactivity. Our mnemonic for these meds is CALMED ADHD.

C, clonidine, is traditionally for resistant hypertension, but what is clonidine's role in ADHD? Clonidine reduces patients' insomnia from other ADHD medicines, or a prescriber may use it alone.

Clonidine is a centrally-acting alpha-2 agonist that reduces sympathetic norepinephrine outflow, decreasing peripheral vascular resistance, and decreasing heart rate and blood pressure. In plain English, it makes it easier for blood to move through the vessels, so heart rate and blood pressure can go down because there's less work to do. Don't stop clonidine abruptly, as rebound hypertension can happen.

You can look at the brand name **Catapres** and think of catabolizing (breaking down) pressure (blood pressure) using the c-a-t-a from catabolizing and p-r-e-s from pressure – **Catapres**.

Prescribers use **clonidine** in A-D-H-D as a single therapy or with stimulants like **methylphenidate,** brand **Concerta**. I had an unusual gym experience with these medicines. A parent had a loud and lengthy talk with a psychiatrist about her child's clonidine and Concerta. While lifting weights, they talked as loud as a rock *concert*. I never forgot the clonidine and Concerta tandem after that. Another brand name for clonidine is **Kapvay**. Take the "p" and "y" out of Kapvay, and you end with K-a-v-a. Kava kava is an herbal that helps calm anxiety, restlessness, and insomnia.

A, atomoxetine, is a non-stimulant for ADHD that inhibits norepinephrine reuptake. While it ends in oxetine, o-x-e-t-i-n-e, it is not an SSRI antidepressant like fluoxetine, brand Prozac. When do we use atomoxetine for ADHD? We use it when stimulants fail or when prescribers have abuse concerns. Unlike stimulants that work immediately, atomoxetine may start working at one week, but not be fully effective until six weeks later. Watch for nausea, increased blood pressure, insomnia, fatigue, and, rarely, liver injury. The brand name **Strattera** uses the s-t-r from <u>str</u>aighten and a-t-t-e from <u>atte</u>ntion, so you can think of straightening a patient's attention, Strattera.

The ADHD stimulants block norepinephrine and dopamine reuptake and are all DEA Schedule II drugs with high potential for abuse and dependence. From stimulants, we expect increased heart rate and blood pressure, insomnia, and decreased appetite. Which patients should avoid ADHD stimulants? Patients suffering from cardiac problems.

L, lis<u>dexamfetamine</u>, comes as a capsule or chewable tablet patients can mix with water or yogurt. The formula has lowered abuse potential for those trying to snort or inject it. Lisdexamfetamine is dextroamphetamine's prodrug.

M, methylphenidate, comes in immediate and extended release formulations. We also see tablet, solution, disintegrating tablet, and patch dosage forms for young children who cannot yet swallow pills. Methylphenidate has many brand names, including **Ritalin** and **Concerta.** Most students already know methylphenidate, but remember that Concerta can help a patient <u>conc</u>ent<u>ra</u>te, taking 'c-o-n-c-e-r' and 't' from concentrate; Concerta. To remember it's an amphetamine, one student of mine said she thought of staying up all night at a <u>concert</u> with Concerta. Concerta is a long-acting once-daily medication.

E,

D, dexmethylphenidate, is methylphenidate's active isomer that comes as a tablet or capsule. Patients can sprinkle it onto applesauce or swallow it whole. Note that we use one-half the methylphenidate dose when dosing dexmethylphenidate. As a student, I remembered **Focalin** and **Concerta** both contain **methylphenidate** as a drug name part. However, I couldn't remember which dexmethylphenidate was. Then I thought of the "F" in "**Focalin**" as following the "d-e" alphabetically in **dexmethylphenidate** to help me. To remember **Focalin** is for ADHD, you can think of **Focalin** helping a patient focus.

A, am<u>phetamine</u>/

D, dextroam<u>phetamine</u>, comes in many dosage forms. Watch for acidic foods and juices that might decrease amphetamine blood levels. The **Adderall** brand reminds us that, once a student can focus, it's no problem to add 'er all up in math class. Another mnemonic is Adderall helps 'ADD' for 'all.'

H,

D, **different formulations, short and long-acting,** allow for the minimum dose. Short-acting formulas help boost parts of a day while long-acting forms cover an entire day. Often dosing coincides with school days.

We'll next move to depression starting with the nine criteria for a depression diagnosis.

NEURO/PSYCH MNEMONIC FLASHCARDS

Question 76. Name the nine criteria for a depression diagnosis.

ESCAPISMS

E_____

S_____

C _____

A _____

P_____

I _____

S_____

M _____

S_____

76. DEPRESSION DIAGNOSIS CRITERIA – ESCAPISMS

> Energy or fatigue
>
> Sleep is hypersomnia or insomnia
>
> Concentration difficulty or indecisiveness
>
> Appetite causing weight loss or gain
>
> Psychomotor agitation or retardation
>
> Interest or pleasure diminished or lost
>
> Suicidal ideations
>
> Mood depressed or irritability in children/adolescents
>
> Self-worth is low, guilt feelings

QUICK SUMMARY

Before we begin talking about antidepressants, let's review what symptoms we expect. Five or more symptoms must come together in a two-week period. Change must happen. For example, if someone slept often, then insomnia would represent change. If they rarely slept, hypersomnia is a variation. Our mnemonic is ESCAPISMS, as many depressed patients want to *escape* their condition.

Next, we'll take a look at SSRIs that can help this condition.

Question 77. Name six SSRIs and how to avoid initial anxiety and agitation as a side effect.

UPS AFFECT

U

P_____ (SSRI)

S_____ (SSRI)

A

Fluo_____ (SSRI)

Fluv_____ (SSRI)

E_____ (left enantiomer SSRI)

C_____ (SSRI)

T_____ (How to avoid initial anxiety and agitation)

77. Selective Serotonin Reuptake Inhibitors (SSRIs) – SSRI UPS AFFECT

U

P<u>aroxetine</u> (Paxil)

S<u>ertraline</u> (Zoloft)

A

F<u>luoxetine</u> (Prozac)

F<u>luvoxamine</u> (Luvox)

E<u>scitalopram</u> (Lexapro)

C<u>italopram</u> (Celexa)

T<u>itrate</u> to avoid agitation and anxiety side effect

Quick Summary

SSRIs block serotonin reuptake, keeping it in the synapse, elevating mood. The UPS AFFECT mnemonic reminds us SSRIs up or raise a depressed patient's affect or mood.

U,

P, paroxetine, has anticholinergic effects including sedation, constipation, and dry mouth, so it's best to give at night. Paroxetine may cause more significant sexual dysfunction than other SSRIs. **Paroxetine** has the **fluoxetine**, "oxetine" stem.

Brand **Paxil** takes "p-a-x"" from **pa**roxetine and "Pax" in Latin means peace. So, looking at Pax + il, the first two letters in illness, Paxil implies peace from the depression illness. The Paxil controlled-release C-R version should have fewer initial side effects and be easier to dose.

S, sertra**line**, is for PMDD or premenstrual dysphoric disorder as well as depression. With insomnia as a side effect, it's better to take sertraline in mornings. **S, ser**tra**line**, has the '-traline, t-r-a-l-i-n-e' stem rather than -oxetine. Brand **Zoloft** "lofts" a depressed patient's mood.

A.

F, fluoxetine, has a long half-life, can cause insomnia, and is also best in the morning. Like sertraline, it's okay for PMDD. We prefer sertraline and fluoxetine in cardiovascular disease. Flu**oxetine** was the first SSRI to make it to market. The "-oxetine, o-x-e-t-i-n-e" ending suggests **fluoxetine**-like entities, but you see "-oxetine" on the S-N-R-I antidepressant dul**oxetine** (brand, **Cymbalta**) and the ADHD medication atom**oxetine** (brand, **Strattera**). Be careful with the -oxetine stem. When fluoxetine gained the premenstrual dysphoric disorder (P-M-D-D) indication, it also gained a new brand name: **Sarafem** combining "Sara, s-a-r-a" like the girl's name and "fem, f-e-m" for feminine. The highest ranked winged angels are Sera-p-h-i-m. Combatting P-M-D-D is the work of angels.

F, fluvoxamine, has another indication, obsessive-compulsive disorder or OCD, while other SSRIs treat various anxiety disorders. Fluvoxamine has more drug interactions inhibiting multiple CYP enzymes.

E, escitalopram, is citalopram's S-enantiomer, so escitalopram's equivalent dose is half citalopram's. While we expect a smaller cardiac toxicity risk than citalopram, the FDA has not changed dosing recommendations. Brand **Lexapro** takes the last four letters of **Celexa**, l-e-x-a, and adds "pro, p-r-o." It's the *professional* upgrade, as **Lexapro** came after **Celexa**. S isomers often enter markets after the original drug loses its patent.

Paroxetine and escitalopram both have FDA approval for anxiety. Can other SSRIs help? Using a drug off-label, or outside FDA approval, is common. Most SSRIs could help, but selecting the right drug is the art of medicine.

C, citalopram, carries a QT prolongation warning with maximum regular doses at 40 mg/day with a 20 mg/day max in patients over 60-years-old. Watch hepatic impairment, and CYP 2C19 medications, bradycardia, hypokalemia, recent MI, and decompensated heart failure. Match citalopram's brand **Celexa** with "relax," as some SSRIs help anxiety also.

T, titrate to avoid agitation and anxiety side effect. Strangely, SSRIs might cause agitation and anxiety when we start them. Let's pause to think about that. If SSRIs can cause anxiety even though they're antidepressant and antianxiety agents, how can we avoid this? Start SSRIs on a low dose and *slowly* titrate to an individualized maintenance dose. Some use the first two S's in SSRI to remind them that you should start S, slowly, slowly.

Side notes:

What non-cardiac SSRI side-effect might stop a patient from taking it and what alternative might help? SSRIs often cause sexual dysfunction. An alternative medication with serotonergic activity which lacks sexual dysfunction is **vilazodone (Viibryd)**, but it has other side effects.

Vilazodone is newer, launched in 2011. It is a selective serotonin reuptake inhibitor with partial agonism at the 5-HT$_{1A}$ receptor. Patients experience fewer sexual side effects than with SSRIs and SNRIs, but it can induce seizures in those with seizure history. Use caution in those patients taking other serotonergic medications due to serotonin syndrome risk.

Next, we'll look at a mnemonic to remember many of these SSRI side effects.

Question 78. Name five serotonin side effects and four serotonin syndrome symptoms.

Side effects - SSRIS

Se _____

S- _____

R _____

I _____

Se _____

Serotonin syndrome symptoms - HARM

H _____

A _____

R _____

M _____

78. SEROTONIN SIDE EFFECTS (SSRIs) AND SEROTONIN SYNDROME SYMPTOMS (HARM)

> **S**exual dysfunction
>
> **S**-isomer
>
> **R**estlessness
>
> **I**nsomnia
>
> **S**erotonin syndrome
>
> **H**igh temperature, hyperthermia
>
> **A**utonomic instability
>
> **R**igidity
>
> **M**yoclonus

QUICK SUMMARY

The S-S-R-I-S part of the mnemonic helps us remember SSRI side effects.

Mnemonic Part 1. SSRIS

S, sexual dysfunction, a significant barrier to compliance.

S, S-isomer of escitalopram has different effects than citalopram.

R, restlessness or CNS stimulation,

I, insomnia. SSRIs, like paroxetine, cause sedation and others like fluoxetine and sertraline cause insomnia. Make sure patients take them at the right time of day.

S, serotonin syndrome which we remember with the HARM mnemonic.

Mnemonic Part 2. HARM

H, hyperthermia;

A, autonomic instability;

R, rigidity; and

M, myoclonus, which includes muscle contractions.

In the next slide, we talk about drugs that might *cause* depression.

Question 79. Name four drugs or drug classes that may *cause* depression.

MOPS ALL DAY

M_____ (antihypertensive)

O_____ C_____ (drug class)

P_____ (antihypertensive)

S_____ (drug class)

79. Drugs causing depression – MOPS ALL DAY

> **M**ethyldopa
> **O**ral contraceptives
> **P**ropranolol
> **S**teroids

Quick summary

Note, cardiovascular disease patients already have an increased depression risk. Consider disease or drug. Depression makes you feel like a dirty mop. Our mnemonic is MOPS ALL DAY.

M, methyldopa, is an antihypertensive.

O, oral contraceptives.

P, propranolol, stands for beta-blockers in general.

S, steroids, can also cause depressive symptoms.

Next, we move to the SNRIs.

Question 80. Name 3 SNRIs, and two unique class suffixes.

TWO HAPPY DVDS

D_____ (SNRI)

V_____ (SNRI)

D_____ (SNRI)

Class suffixes:

-o_____

-f_____

80. SEROTONIN NOREPINEPHRINE REUPTAKE INHIBITORS (SNRIs) – TWO HAPPY DVDS

> **D**ul*oxetine* (Cymbalta)
>
> **V**enla*faxine* (Effexor)
>
> **D**esvenla*faxine* (Pristiq)

QUICK SUMMARY

What common side effects should we expect from SNRIs based on the neurotransmitters they affect? SNRIs' serotonin can cause drowsiness *or* insomnia. Adding norepinephrine adds increased heart rate, dry mouth, sweating, and constipation. SNRIs may make patients feel worse before they feel better.

The SNRI mnemonic of TWO HAPPY DVDs refers to TWO neurotransmitters S, serotonin and N, norepinephrine, HAPPY, the therapeutic result, and DVDs which stands for the three SNRI generic drug names' first letters.

D, dul*oxetine*, works for depression but also has neuropathy, fibromyalgia, and generalized anxiety indications. Watch for orthostasis. Duloxetine ends in "oxetine," but it's an SNRI, not SSRI. Duloxetine affects serotonin *and* norepinephrine, so think of the "du" as a duo for two. We pronounce d-u, du like French deux, d-e-u-x, which also means two. I have never seen an unhappy cymbal player in a band, and "alta" means tall in Spanish. So, I picture a happy, tall cymbal player to remember that brand **Cymbalta** elevates mood.

V, venla<u>faxine</u>, favors serotonin reuptake over norepinephrine reuptake. At high doses, however, the norepinephrine reuptake effects may increase blood pressure. Venla<u>faxine</u> treats depression *and* anxiety. If you look at the "afax, a-f-a-x" in **venla<u>faxine</u>** and "Effex, E-f-f-e-x" in brand name **<u>Effex</u>or,** you can see it "effects" both.

D, desvenla<u>faxine</u>, venlafaxine's enantiomer, only has a depression indication with similar side effects. Both have the "-faxine, f-a-x-i-n-e," stem.

We follow up SNRIs with TCA antidepressants.

Memorizing Pharmacology Mnemonics

Question 81. Name 5 TCAs and 1 TeCA.

WITHOUT DOPAMINE

DO_____ (TCA, tertiary amine)

P_____ (TCA, secondary amine)

A_____ (TCA, tertiary amine)

M_____ (TeCA)

I_____ (TCA, tertiary amine)

N_____ (TCA, secondary amine)

E

81. Tricyclic and Tetracyclic Antidepressants (TCAs, TeCAs) – WITHOUT DOPAMINE

> **Do**xepin (Silenor)
> **P**rotriptyline (Pamelor)
> **A**mitriptyline (Elavil)
> **M**irtazapine (Remeron) (a TeCA)
> **I**mipramine (Tofranil)
> **N**ortriptyline (Pamelor)
> **E**

Quick summary

Tricyclic antidepressants (TCAs), also treat depression. Like SNRIs, they inhibit serotonin and norepinephrine reuptake, but block acetylcholine and histamine worsening their side effect profile. TCAs chemically, have three rings making them a *tri*cyclic, but mirtazapine, a *tetra*cyclic, has four rings. The mnemonic WITHOUT DOPAMINE highlights these antidepressants lack of effect on the dopamine neurotransmitter.

DO, doxepin, like amitriptyline, has a higher anticholinergic side effect risk. Doxepin treats depression, insomnia, and works for itching as a cream. Think of beating depression as getting one's ducks in a row. Picture "ducks, d-u-c-k-s" and duck pin, p-i-n – bowling related to dox-e-pin, **doxepin**. The brand

Silenor has every letter from snore, s-n-o-r-e or the first five letters of silent, s-i-l-e-n-t. Silent snore Silenor.

P, protr**iptyline** has ability to "p," promote wakefulness and help with "p," pain like migraines. Sleeplessness contrasts most TCAs which cause drowsiness.

A, amitr**iptyline**, is for depression, neuropathic pain, and migraine prophylaxis. As a tertiary amine, it includes more side effects like dry mouth, constipation, blurred vision, sedation, and weight gain. Use caution in the elderly with these anticholinergic effects.

Amitriptyline's stem, is "triptyline, t-r-i-p-t-y-l-i-n-e." The t-r-i-p, then t-y-l, reminds us it "trips" up depression "until" it's gone. "Tri" in **ami**tr**iptyline** implies a *tri*cyclic antidepressant. Brand **Elavil** elevates the patient's mood.

M, mirtazapine, is a tetracyclic antidepressant. It is an alpha-2 antagonist, which increases norepinephrine and serotonin release. By increasing appetite, it helps elderly patients who struggle to gain weight. Mirtazapine has anticholinergic side effects like dry mouth and constipation. Remember mirtazapine's sedation side effect from its brand **Remeron** which has REM, R-E-M, suggesting rapid-eye-movement sleep.

I, imipramine.

N, nortr**iptyline** is a secondary amine with fewer side effects than tertiary amines.

E.

We've covered a lot of material in the neuro chapter, let's do five quick review questions to mix it up.

1) **TCA and BZDs are *second-line* treatments for panic attacks; why?** Undesirable side effects. Benzos have abuse potential; we use TCAs only after safer SSRI failure.
2) **Are SSRIs or TCAs first-line therapy for obsessive-compulsive disorder, OCD, and why?** SSRIs, because they are safer.
3) **Are norepinephrine, serotonin, and dopamine part of the central nervous system or the peripheral nervous system?** Central nervous system.
4) **How do SSRIs and SNRIs work? Do they increase or decrease synaptic neurotransmitter levels?** They block serotonin and/or norepinephrine reuptake, increasing the neurotransmitters available to activate receptors.
5) **Will SNRIs cause hypertension or hypotension and why?** SNRIs increase norepinephrine, activating receptors and causing increased blood pressure causing hypertension.

Next, we look at TCA side-effects.

Memorizing Pharmacology Mnemonics

Question 82. Name four side effects or interactions with TCAs.

TCAS

T_____ (side effect)

C_____ (caution, Beers Criteria)

A_____ (side effect)

S_____ (caution)

82. Tricyclic antidepressants - TCAS

> **T**ired, we use some as sedatives
>
> **C**aution with falls from orthostatic hypotension, confusion, and Beers Criteria.
>
> **A**nticholinergic effects, use the ABDUCT mnemonic
>
> **S**uicide risk, watch for overdose

Quick summary

The acronym for tricyclic antidepressants, TCAS, serves as our mnemonic.

T, tired. We use some as sedatives. Most TCAs cause sedation but remember, protriptyline *pro*motes wakefulness.

C, caution with falls from orthostatic hypotension, confusion, and Beers Criteria.

A, anticholinergic effects, use the ABDUCT mnemonic to remember the drying effects to expect.

S, suicide risk, watch for overdose because it only takes about a week's worth of TCAs to threaten someone's life. Don't use in suicidal patients or supervise the daily doses.

Our final antidepressant class includes the monoamine oxidase inhibitors, MAOIs, next.

Question 83. Name four MAOIs and two either dietary or drug combination restrictions.

PINOTS

P_____ (MAOI)

I_____ (MAOI)

N_____ from tyramine in cheese, chocolate

O_____ without extra S_____

T_____ (MAOI)

S_____ (MAOI)

83. MONOAMINE OXIDASE INHIBITORS (MAOIs) – PINOTS

> **P**henelzine (Nardil)
>
> **I**socarboxazid (Marplan)
>
> **N**orepinephrine from tyramine in cheese, chocolate, etc.
>
> **O**nly without extra serotonin
>
> **T**ranylcypromine (Parnate)
>
> **S**elegiline (Eldepryl, Emsam) (Type B)

QUICK SUMMARY

MAOIs block monoamine oxidase, an enzyme, which typically breaks down serotonin, norepinephrine, and dopamine. If we inhibit its breakdown, we increase neurotransmitter levels which are good for depression. However, side effects and food and drug interactions that decrease efficacy make this therapy third-line. The mnemonic PINOT, P-I-N-O-T reminds us of red wine. Red wine contains tyramine, which interacts with MAOIs to cause a potential hypertensive crisis.

P, phenelzine, is more sedating and patients should take at night.

I, isocarboxazid. A student came up with the atypical sad man who laments, "I so carve boxes" for **isocarboxazid**. Additionally, you can take the "m" and "a" from isocarboxazid's **Marplan** to remember it's an M-A-O-I.

N, norepinephrine, from tyramine in cheese, chocolate, etc. is dangerous because the gut usually destroys norepinephrine with monoamine oxidase. With an inhibitor, we see an increase. We find tyramine in aged cheeses, cured meats, tap beer, and soy sauce. Excess norepinephrine increases blood pressure. Eating tyramine-rich foods with MAOIs may lead to hypertensive crisis.

O, only without extra serotonin. MAOIs, block our body's ability to break down serotonin, increasing serotonin and risk of serotonin syndrome. As such, we don't give MAOIs with SSRIs, SNRIs, TCAs, or bupropion. Serotonin syndrome presents a mental status change triad, neuromuscular change, including tremor and myoclonus, and autonomic dysregulation, including tachycardia.

T, tranylcypromine, is stimulating. Give in the morning.

S, selegiline's s-e-l-e can remind you that it's a *selective* irreversible MAO-B inhibitor. Its brand **Eldepryl** breaks down as "elderly pill" as Parkinson's disease primarily affects people 60 years and older. Note, only selegiline patches, brand **Ensam**, are for depression. In the tablet form, selegiline inhibits MAO-B, which does not help depression.

We'll look at an alternative MAOI mnemonic next.

Question 84. Name four MAOIs.

THE PITS

P_____ (MAOI)

I_____ (MAOI)

T_____ (MAOI)

S_____ (MAOI Type-B)

84. Monoamine oxidase inhibitors for depression - THE PITS

> **P**henelzine (Nardil)
> **I**socarboxazid (Marplan)
> **T**ranylcypromine (Parnate)
> **S**elegiline (Eldepryl, Emsam) (Type B)

Quick Summary

Depression is THE PITS is an alternate way to remember the four monoamine oxidase inhibitors, MAOIs.

From antidepressants, we'll move to mood stabilizers.

Question 85. Name four mood stabilizers and which is an autoinducer.

VOCALL MOOD

V_____

O

C_____ which is an A_____

Li_____

La_____

85. Mood Stabilizers OCD – VOCALL MOOD

> **V**alproic acid (Depakote)
> **O**
> **C**arbamaze<u>pine</u> (Tegretol)
> **A**utoinducer (carbamazepine)
> **L**ithium (Lithobid)
> **L**amo<u>trigine</u> (Lamictal)

Quick summary

Bipolar may include extreme emotional highs and hyperactivity called mania, emotional lows as depression, or signs and symptoms of both. Patients with bipolar disorder, in the manic state, may talk loud and fast, so our mnemonic is VOCALL, with two L's.

V, valproic acid, or valproate or divalproex, can increase liver function tests (LFTs) and cause fetal harm. It can also rarely cause a rash, high ammonia levels, and thrombocytopenia, putting patients at bleeding risk. Valproic acid has many drug-drug interactions with other antiepileptics. While you find "v-a-l" in many medication names, think of the "val" in di<u>val</u>proex as "vul, v-u-l" in con<u>vul</u>sions. One student thought of dival-pro-ex as a <u>pro</u> at <u>ex</u>tracting seizures – divalproex.

O.

C, carbamazepine, is a mood stabilizer and an antiepileptic that can cause severe skin reactions like Stevens-Johnson syndrome, especially in those of Asian descent who possess the HLA-B*1502 allele. Aplastic anemia and agranulocytosis are rare side effects. While carbamazepine's "-pine," [[pronounced, "peen,"]] p-i-n-e is the stem, it's not very useful because the "-pine" ending means a chemical has three rings. Instead, think either of seizures being "<u>carb</u>ed" (curbed) or of being "<u>amaze</u>d" that the seizures are "sto-<u>pp</u>i<u>n</u>g" – using the letters inside **carb-amaze-pine**. Brand **Tegretol** has the scattered letters "t-r-o-l" in con<u>trol</u>, as in to control seizures: **Tegretol**.

A, autoinducer. Carbamazepine is an autoinducer, inducing its own metabolism. Shortly after a patient starts therapy, the drug's half-life decreases, leading to lower serum blood levels.

L, lithium. Unlike carbamazepine and valproic acid, which are also for epilepsy, lithium is only for bipolar disorder. It is old, effective, and helps prevent suicide. Lithium side effects we'll see in two mnemonics forward. The "b-i-d" in the brand name **Lithobid** is the Latin <u>b</u>is-<u>i</u>n-<u>d</u>ie, or twice-daily for the dosing schedule.

L, lamotrigine, is an FDA approved antiepileptic for bipolar. It's most useful for depressive episodes, but it does so without significantly precipitating manic episodes

Next, we'll look specifically at some of these medications' side effects and the mechanism of carbamazepine.

Memorizing Pharmacology Mnemonics

Question 86. Name five important properties of carbamazepine.

CARBA

C_____

A_____

R_____ S_____ (antiepileptic)

B_____ disorder treatment

A L_____ T_____ to stead state

86. Carbamazepine - CARBA

> **C**lassify as a traditional agent
>
> **A**utoinduces metabolism to decrease the concentration
>
> **R**educe seizures as an antiepileptic
>
> **B**ipolar disorder treatment
>
> **A** long time to steady state.

Quick Summary

Our mnemonic is CARBA, the first five letters in carbamazepine.

C, classify as a traditional agent as opposed to newer agents like gabapentin and pregabalin.

A, autoinduces metabolism to decrease concentration, an important consideration with dosing.

R, reduce seizures as an antiepileptic.

B, bipolar disorder treatment.

A, a long time to steady state. It may take a month for full effectiveness.

Next, we have lithium specifics for bipolar disorder.

Memorizing Pharmacology Mnemonics

Question 87. Name seven side effects of lithium.

LITHIUM

L_____ (side effect)

I_____ or group I in Periodic Table

T_____ (side effect)

H_____ (side effect)

I_____ (side effect)

U_____ (tremors)

M_____ L_____ (method of assessment)

87. LITHIUM: SIDE EFFECTS - LITHIUM

> **L**eukocytosis
>
> **I**on or Group I in Periodic Table of Elements
>
> **T**eratogen
>
> **H**ypothyroid
>
> **I**ncreased urine
>
> **U**nstable (tremors)
>
> **M**onitor levels

QUICK SUMMARY

Our mnemonic is the drug name itself, LITHIUM.

L, leukocytosis is a high white blood cell count usually signaling infection or inflammation.

I, ion or Group I in Periodic Table of Elements. Lithium sits in the same group on the far left of the Periodic Table of Elements as the Latin *Natrium* (N-a), commonly known as the chemical element sodium. Because of this, patients taking lithium must roughly consume the same amount of sodium regularly. Watch for changes in renal function because of high renal clearance. Therefore, "Where the salt goeth, the **lithium** goeth too."

T, teratogen, harmful to a fetus.

H, hypothyroid. Lithium creates a reversible inhibition of thyroid hormone release.

I, increased urine also known as polyuria

U, unstable, can cause tremors.

M, monitor levels. Lithium has a narrow therapeutic window. The helpful and harmful dose are close together requiring tests and tight control. Watch for issues with renal elimination.

Next, we cover the first-generation antipsychotics.

NEURO/PSYCH MNEMONIC FLASHCARDS

Question 88. Name two 1st generation antipsychotics in order from high to low potency, a drug class that can help movement disorders, and two side effects we might expect with this class.

NO POTENT CHATS

C_____ (1st generation antipsychotic)

H_____ (1st generation antipsychotic)

A_____ (drug class to help with movement disorders)

T_____ D_____ (side effect)

S_____ (side effect)

88. 1ST GENERATION ANTIPSYCHOTICS – NO POTENT CHATS

> **C**hlorpromazine (Thorazine) (Low potency, sedation)
> **H**aloperidol (Haldol) (High potency, EPS)
> **A**nticholinergics help with some movement disorders
> **T**ardive dyskinesia and other extrapyramidal symptoms
> **S**edation

QUICK SUMMARY

All first-generation antipsychotics have the risk of causing extrapyramidal symptoms, EPS, or movement disorders. Second-generation antipsychotics are not more effective – instead, they have different side effects, like metabolic side effects with less risk of EPS.

We further divide the drugs as low or high potency, which coincides with their level of dopamine blockade. Low potency, in plain terms, means that it will take about 100 milligrams of chlorpromazine to do what 2 milligrams of high potency haloperidol does. Fewer milligrams to get the same work done is higher potency.

Schizophrenia involves psychosis, reality disturbances, and perceptual misinterpretation. We often use inexpensive first-generation or typical antipsychotics for positive symptoms. The CHATS mnemonic refers to voices in schizophrenia, a positive symptom of hallucinations or delusions. Negative symptoms include social withdrawal, lack of motivation, and poor self-care. Cognitive symptoms include difficulty planning and making decisions.

C, chlorpromazine, a low-potency first-generation antipsychotic affects dopamine less, and patients likely have fewer extrapyramidal symptoms, EPS, but more sedation. Chlorpromazine was the first antipsychotic. While it carries side effects, it represented a new treatment option for schizophrenic patients – a first generation. Generational classifications are especially crucial in antipsychotics because of differences in side effects. Thor is a mythical god, and you can think of chlorpromazine's brand **Thorazine** as helping people who have delusions of mythical people.

H, haloperidol, a high potency first-generation antipsychotic affects dopamine more, leading to greater extrapyramidal symptoms or EPS. When a first-generation antipsychotic is high potency, it works more on dopamine than low potency, and it leads to more extrapyramidal symptoms, or EPS, which is a group of four movement disorders: akathisia, dystonia, pseudoparkinsonism, and tardive dyskinesia. Watch for QTc prolongation as well. Use the "-peridol, p-e-r-i-d-o-l" stem to recognize this first-generation high potency drug. Many students think of the "halo" in haloperidol as the halo that sits *high* on an angel's head to remember this is *high* potency. To make the brand name **Haldol**, they just took the first and last three letters of the generic name haloperidol; Haldol. The haloperidol decanoate form, a long-acting injection, we remember with "decanoate lasts a decade."

A, anticholinergics help with some movement disorders.

T, tardive dyskinesia, we associate with abnormal facial movements, especially in the mouth or tongue. It can be irreversible. However, a drug came out in 2017 called **valbenazine tosylate (Ingrezza)**, which has some effectiveness for tardive dyskinesia.

S, sedation, typical to chlorpromazine, a low-potency first-generation antipsychotic.

From first-generation, we move to second-generation antipsychotics.

Question 89. Name the four significant extrapyramidal symptoms (EPS) brought on primarily by first-generation antipsychotics and what someone might use for early EPS.

NEVER ADAPT

A_____ (side effect)

D_____ (side effect)

A_____ (drug class for early EPS)

P_____ (side effect)

T_____ (side effect)

89. Extrapyramidal symptoms (EPS) – Never Adapt

> **A**kathisia
>
> **D**ystonia
>
> **A**nticholinergics for early EPS
>
> **P**arkinsonism
>
> **T**ardive dyskinesia

Quick Summary

Extrapyramidal symptoms come from some antipsychotics. Some are irreversible, and patients may NEVER ADAPT.

A, akathisia is an inability to remain still. A<u>kin</u>esia, is the absence of movement. Think a + kinesia from kinesiology, like the college major, makes it 'not' + 'movement.'

D, dystonia includes prolonged muscle contractions, which can occur anywhere from the eyes to jaw and tongue.

A, anticholinergics for early EPS, benztropine for example.

P, pseudoparkinsonism includes rigidity, slow movement, and tremors.

T, tardive dyskinesia includes possibly irreversible abnormal facial movements, especially in the mouth or tongue.

From extrapyramidal symptoms, we move to second-generation antipsychotics which minimize this effect.

Question 90. Name eight 2nd generation antipsychotics naming five class suffixes.

ABC OPQR & Z

A_____ with suffix -p_____

B_____ with suffix -p_____

C_____ with suffix -p_____

O_____ with suffix -p_____

P_____ with suffix -p_____

Q_____ with suffix -t_____

R_____ with suffix -p_____

&

Z_____ with suffix -s_____

90. 2ND GENERATION ANTIPSYCHOTICS – ABC OPQR & Z

> **A**ripiprazole (Abilify)
> **B**rexpiprazole (Rexulti)
> **C**lozapine (Clozaril)
>
> **O**lanzapine (Zyprexa)
> **P**aliperidone (Invega)
> **Q**uetiapine (Seroquel)
> **R**isperidone (Risperdal)
>
> and
>
> **Z**iprasidone (Geodon)

QUICK SUMMARY

Second-generation antipsychotics (SGAs) have some similarities with first-generation antipsychotics (FGAs), but they are different.

We more commonly employ second-generation or atypical antipsychotics which have fewer extrapyramidal side effects (EPS) than their first-generation counterparts. However, second-generation drugs have more metabolic side effects like weight gain, dyslipidemia, and glucose dysregulation. The

ABC, OPQR, and Z alphabetic mnemonic order helps us keep this alphabet soup in our heads.

A, aripiprazole, and

B, brexpiprazole, are both second-generation antipsychotics with the '-piprazole, p-i-p-r-a-z-o-l-e' stem. While -piprazoles have some metabolic effects, like glucose and lipid dysfunction, they can occasionally cause akathisia or constant restlessness. Aripiprazole has a risk of QTc prolongation and is available as a long-acting injection.

The World Health Organization (WHO) discourages the "–piprazole" stem because it has the PPI "–prazole" stem in it. Don't confuse them with '-prazole, p-r-a-z-o-l-e' from proton pump inhibitors, PPIs.

Aripiprazole's **Abilify** helps a patient with schizophrenia have more "ability to function in society." Brexit was Britain's break from the U.K. **Brexpiprazole** is for schizophrenia, with 'schism' meaning break. Brexpiprazole's brand **Rexulti** gets 'results.'

C, clozapine is one of the most effective antipsychotics available, but with many side effects, we reserve it for treatment failure cases. Clozapine has lower suicide rates, but we watch clozapine "clozely, c-l-o-z-e-l-y" for agranulocytosis, seizures, and anticholinergic side effects. **Agranulocytosis** is a reduced white blood cell count which results in an inability to fight infections. Low white blood cell counts might force therapy to stop. Therapy may not restart if the count drops too low. Dose-related **seizures** can occur with rapid titration. **Anticholinergic** side effects like dry mouth, constipation, and blurred vision are common.

O, olanzapine, has a high incidence of metabolic effects, including weight gain, dyslipidemia, and glucose dysregulation. Some use the "O" in olanzapine for "obesity" as a side effect. Olanzapine does not have the severe clozapine side effects and comes as a long-acting injectable.

P, paliperidone.

Q, quetiapine. Quetiapine has a low risk of EPS, moderate weight gain effects and a higher chance for dyslipidemia. Patients might feel dizzy and sleepy. If you switch the "i" and "t" in **quetiapine**, q-u-e-t-i-a-p-i-n-e, you get the word "quiet," as in quieting the voices. **Seroquel**, the brand name, shares the "q-u-e" from **quetiapine**, and quell means to silence someone.

R, risperidone has a Parkinsonism risk at higher doses with mild metabolic side effects. Some use risperidone's "per, p-e-r" to represent the pair of breasts that may enlarge in men. Excess prolactin shows as lactation and sexual dysfunction in women. Note the stem "–peridol, p-e-r-i-d-o-l" from **haloperidol** and "–peridone, p-e-r-i-d-o-n-e" from **risperidone** are similar; and both have long-acting injectable forms. Brand **Risperdal** and generic **risperidone** share the first letters "r-i-s-p-e-r,"risper. Think of risper and whisper, as in calming the whispering voices.

Z, ziprasidone.

From antipsychotics, we'll move to antiepileptics that help seizure patients.

Question 91. Name three traditional anti-epileptics.

PACED

P_____

A

C_____

E

D_____

Memorizing Pharmacology Mnemonics

91. Traditional Anti-Epileptics – PACED

> **P**henytoin (Dilantin)
> **A**
> **C**arbamazepine (Tegretol)
> **E**
> **D**ivalproex (Depakote)

Quick summary

These traditional anti-epileptics for local and generalized seizures regulate neuronal action potentials preventing attacks. There are two antiepileptic types – traditional or older agents and non-traditional or newer agents.

Since the older agents have been around longer, we know their efficacy in different seizure types. Older antiepileptics, unfortunately, are less well tolerated, have worse side effect profiles, and complex pharmacokinetics affecting absorption and distribution. We use the PACED mnemonic to remember motion, as a person with epilepsy might have.

P, phenytoin, is unlike carbamazepine and divalproex, which have bipolar disorder indications. Phenytoin solely controls seizures. Like lithium, phenytoin has a narrow therapeutic index, so small dosage differences can mean toxicity.

Phenytoin exhibits saturable kinetics. What does that mean? Typically, as we increase the medication dose, we increase the concentration linearly which manifests as a diagonal line going from bottom left to top right on a concentration versus time chart. With phenytoin, small dose changes can lead to exponential concentration increases, so we monitor levels closely. On a graph, you'd see the right half of the letter 'U.'

Side effects include rash, agranulocytosis, and gingival hyperplasia. Acute phenytoin toxicity presents as nystagmus, or uncontrolled eye movements, blurred vision, and ataxia, a lack of muscle control. The "toin, t-o-i-n" stem helps you remember phenytoin is an antiepileptic. **Dilantin** and shakin' also rhyme.

A,

C, **carbama**ze**pine**, is a mood stabilizer and antiepileptic we've discussed earlier.

E,

D, divalproex, we also discussed earlier.

Next, we move to phenytoin side effects.

Question 92. Name nine possible adverse effects of phenytoin.

PHENYTOIN

P-_____ interactions

H_____

E_____

N_____

Y_____ - B_____ of the skin

T_____

O_____

I_____ with B-12 metabolism

N_____

   Example 1. V_____

   Example 2. A_____

   Example 3. H_____

92. Phenytoin: Adverse Effects
PHENYTOIN

> **P**- 450 interactions
>
> **H**irsutism, hair growth
>
> **E**nlarged gums, gingival hyperplasia
>
> **N**ystagmus
>
> **Y**ellow-browning of skin
>
> **T**eratogenic
>
> **O**steomalacia. Original kinetics (Michaelis-Menten)
>
> **I**nterferes with B-12 metabolism (hence anemia)
>
> **N**europathies: vertigo, ataxia, headache

Quick Summary

While these side effects are important, watching for sedation and cognitive impairments in the elderly and children is crucial.

Next, we'll cover newer antiepileptics, which are not always better than traditional anti-seizure medicines.

Question 93. Name eight newer antiepileptics.

LAPTOP LIGHTZ

L_____

A

P_____

T_____

O_____

P

L_____

I

G_____

H

T_____

Z_____

93. Newer Antiepileptics – LAPTOP LIGHTZ

> **L**amo<u>trigine</u> (Lamictal)
>
> **A**
>
> **P**re<u>ga</u>balin (Lyrica)
>
> **T**opiramate (Topamax)
>
> **O**xcarbaze<u>pine</u> (Trileptal)
>
> **P**
>
> **L**eveti<u>racetam</u> (Keppra)
>
> **I**
>
> **G**abapentin (Neurontin)
>
> **H**
>
> **T**ia<u>ga</u>bine (Gabitril)
>
> **Z**onisamide (Zonegran)

Quick Summary

In general, patients tolerate newer antiepileptics better. They have fewer drug-drug interactions and are safer when it comes to pregnancy. Unfortunately, newer antiepileptics haven't been around as long and have less proved places in therapy. We use a slow titration schedule for most. Lights can worsen seizures, so we use the mnemonic LAPTOP LIGHTZ.

L, lamotrigine, is a sodium channel blocker that works on all types of seizures. It can cause dizziness, nausea and rarely a rash that can progress to Stevens-Johnson syndrome, especially with rapid dose increases with initial titration. Lamotrigine's brand name **Lamictal** has "ictal," in it, meaning seizure

A,

P, pregabalin, like gabapentin, blocks calcium channels. Like gabapentin, it is for neuropathy, postherpetic neuralgia, and fibromyalgia. It can cause peripheral edema and weight gain as well as euphoria. Pregabalin is a controlled substance, DEA Schedule V. The "gab, g-a-b" stem is a little misleading. Neither **pregabalin** nor **gabapentin** directly affects gamma-amino-butyric-acid (GABA) receptors. A lyre is a musical instrument. A lyric is a song line. Either can remind you of a stopped seizure in harmony with pregabalin's **Lyrica**.

T, topiramate, is a sodium channel blocker. Side effects limit its use. CNS side effects include sedation, fatigue, and concentration problems. Other rare side effects include kidney stones, metabolic acidosis, and hyperammonemia and reduced sweating. It can cause weight loss. In both the generic and brand names, topiramate and **Topamax**, you can use the t-o-p as in s-t-o-p, to stop seizures. Topamax sounds like Dopamax, and topiramate makes you dopey.

O, oxcarbazepine, is a sodium and calcium channel blocker with a similar structure to carbamazepine without autoinduction. It can cause sedation, dizziness, and nausea, with a higher risk of producing low sodium than carbamazepine. A rash can occur with oxcarbazepine, so don't use it in patients allergic to carbamazepine. By changing the "carb" to "curb," you can think of curbing seizures with **oxcarbazepine**. The brand **Trileptal** has the e-p-i-l from epileptic.

P,

L, levetiracetam, blocks sodium channels, increases GABA, and works for multiple seizure types. Patients tolerate it well, with side effects including dizziness, weakness, and drowsiness. It can (rarely) cause behavioral changes like irritability, anxiety, or depression. In **levetiracetam's** brand **Keppra**, you can use the K-e-p, Kep and p-r-a, pra to think that an epileptic "kept praying" the seizures would stop.

I,

G, gabapentin, inhibits calcium channels. The 'gab' prefix in gabapentin is misleading because while we don't understand the mechanism of action, there is no effect on GABA. Gabapentin isn't useful for seizures, but we often use it for neuropathy and postherpetic neuralgia. Side effects include dizziness, drowsiness, edema, and weight gain. Absorption is saturable, so bioavailability at higher doses decreases. The "neu, n-e-u" [[pronounce "new"] in **Neurontin** is one way to remember that this is a newer drug.

T, tiagabine, is also not very effective, as it can cause new-onset seizures and epilepsy in patients without epilepsy. Use sparingly and don't use off-label.

Z, zonisamide.

Next, we turn to drugs used to treat Parkinson's.

Memorizing Pharmacology Mnemonics

Question 94. Identify eight medications associated with Parkinson's treatment.

SUBPAR CLUES TO CAUSE

S_____

U

B_____

P_____

A_____

R_____

C_____/L_____

U

E_____

S_____

94. Parkinson's Agents – SUBPAR CLUES TO CAUSE

> **S**elegiline (Eldepryl)
> **U**
> **B**enztropine (Cogentin)
> **P**ramipexole (Mirapex)
> **A**mantadine (Gocovri)
> **R**opinirole (Requip)
>
> **C**arbidopa /
> **L**evodopa (Sinemet)
> **U**
> **E**ntacapone (Comtan)
> **S**afinamide mesylate (Xadago)

Quick Summary

Motor symptoms like slow movement, tremors, and rigidity characterize Parkinson's disease. A condition related to having too little dopamine, we work to increase dopamine as treatment. Our mnemonic SUBPAR CLUES TO CAUSE reminds us that we aren't sure why Parkinson's responds to medicine, but later, the drug stops working.

S, selegiline, is a monoamine oxidase-B inhibitor, or MAO-B inhibitor. We talked about selegiline before with antidepressants, where monoamine oxidase degrades neurotransmitters, so we give the inhibitor to increase neurotransmitter levels. By inhibiting MAO-B, we prevent dopamine metabolism. However, the MAO inhibition increases the serotonin syndrome potential with other serotonin agents. The brand name **Eldepryl** helps you remember that it relieves symptoms of Parkinson's disease, a condition more prevalent in the "elderly," Eldepryl.

U,

B, benztropine, is an anticholinergic medication for Parkinson's tremors or, with antipsychotics, for Parkinsonism. This anticholinergic's side effects include constipation, blurred vision, and dry mouth. Prescribers use this cautiously in the elderly, as it can cause confusion and an increased risk for falls. The brand name **Cogentin** hints at cognition.

P, pramipexole, is a dopamine agonist I'll review with ropinirole. The brand name **Mirapex** twists around the generic name, pramipexole.

A, amantadine, interestingly, can treat influenza and Parkinson's. In brand **Gocovri** you can find many letters of recover, as in "recover from the disease."

R, ropinirole, and pramipexole are dopamine agonists that act like dopamine at the dopamine receptor. Side effects include drowsiness, sleep attacks, hallucinations, and (rarely) impulse control behaviors. Both have a role in restless leg syndrome. Ropinirole's brand **Requip** "equips" a patient to deal with Parkinson's.

C, carbidopa, and **L, levodopa**, increase available dopamine. Levo<u>dopa</u> can work in the brain but can also float in the periphery, so carbi<u>dopa</u> prevents dopamine breakdown there, leading to fewer systemic side effects and more dopamine to the brain. Side effects include nausea and dizziness. Long-term use might lead to dyskinesias and the drug losing its effectiveness. The stem "dopa, d-o-p-a" in both levodopa and carbidopa help remind us that increased dopamine is critical. The brand *Sinemet* combines levo<u>dopa</u> and carbi<u>dopa</u> to work *syn*ergistically. Carbi<u>dopa</u> doesn't have an antiparkinsonian effect. Rather, it reduces levodopa breakdown to increase the availability to the patient.

If levodopa treats PD, a movement disorder, then why does it cause dyskinesias? Levodopa has a narrow therapeutic window that changes over time. Too much or little dopamine causes dyskinesias. You need a perfect balance. Dyskinesias occur before or soon after the "on-off" phenomenon. We can reduce the dose, but this worsens Parkinson's symptoms.

U,

E, entacapone, is a COMT, catechol-O-methyl transferase inhibitor. It's like carbidopa in that it prevents levodopa breakdown, extending its duration of action. Use to combat the "wearing off" levodopa effect.

S, safinamide mesylate, is a Parkinson's add-on to levodopa or other medications, also for the levodopa "wearing off."

While Parkinson's involves a shortage in dopamine, we'll next cover Alzheimer's, a lack of acetylcholine.

Question 95. Name four drugs for Alzheimer's and the neurotransmitter in short supply in this condition.

DREAMY GAL

D_____ (acetylcholinesterase inhibitor)

R_____ (acetylcholinesterase inhibitor)

E

A_____ (neurotransmitter)

M_____ (NMDA receptor antagonist)

Y

GAL_____ (acetylcholinesterase inhibitor)

95. Alzheimer's Agents – DREAMY GAL

> **D**onepezil (Aricept)
> **R**ivastigmine (Exelon)
> **E**
> **A**cetylcholine
> **M**emantine (Namenda)
> **Y**
>
> **GAL**antamine (Razadyne)

Quick Summary

When patients have Alzheimer's, they sometimes feel like they are in a dream, and can't remember, so our mnemonic is DREAMY GAL.

D, donepezil, is a cholinesterase inhibitor blocking acetylcholine breakdown by the acetylcholinesterase enzyme leaving more in the brain. However, cholinesterase inhibitors like donepezil commonly cause CNS side effects like dizziness and insomnia, as well as nausea, diarrhea, and urinary incontinence. Donepezil is for mild, moderate, and severe Alzheimer's disease. When I think of **donepezil** as an Alzheimer's medication, I think, "I don't remember zilch!" – taking "d-o-n" from donepezil's front and "z-i-l" from the back. The brand **Aricept** improves perception and Alzheimer's patients' powers of recollection.

R, rivastigmine, comes in a patch formulation.

E,

A, acetylcholine, is the neurotransmitter we're trying to increase in the brain.

M, memantine, is an NMDA receptor antagonist and an investigational drug for autism. In Alzheimer's, it's thought there is an overstimulation of glutamate on NMDA receptors, so if we block this, we can protect against further damage. Side effects include constipation and dizziness. Memantine is only for moderate to severe Alzheimer's disease. The generic name memantine has "mem, m-e-m" for memory in it. The brand name **Namenda** comes from N-Methyl-D-aspartate (N-M-D-A), the receptor it antagonizes. Think Namenda as mends the brain.

Y,

GAL, galantamine.

Next, we'll move to look at local anesthetics.

Question 96. Identify three local anesthetics and the class suffix.

BLOCK AXONS

B_____ (local anesthetic, ester)

L_____ (local anesthetic, amide)

O

C_____ (DEA scheduled local anesthetic)

K

Local anesthetics class suffix: -c_____

96. LOCAL ANESTHETICS – BLOCK AXONS

> **B**enzocaine (Anbesol)
> **L**idocaine (Xylocaine)
> **O**
> **C**ocaine
> **K**

Quick Summary

Local anesthetics block action potentials traveling down an axon, so they prevent pain signals from going to and from your brain. The mnemonic for local anesthetics is BLOCK, as local anesthetics *block* nerve impulse conduction.

B, benzocaine, has the "caine, c-a-i-n-e" stem that points out it's a local anesthetic. **Anbesol** has the letter "n," and the letter "b" in it, the first and last letter of the word numbs for its use in numbing an aching tooth.

L, lidocaine, differs slightly based its chemical shape. Benzocaine is an ester-type anesthetic, and lidocaine is an amide-type anesthetic. Ester anesthetics are more allergenic. Therefore, we use them as topicals. Amide anesthetics like lidocaine are safer than esters as injections. There are cardiovascular indications as well in the emergency drugs section. We often use lidocaine topically over-the-counter to treat sunburns, therefore the brand name **Solarcaine.** Injectable and patch forms of lidocaine are also available by prescription. In emergencies, paramedics often use lidocaine for arrhythmias as an injectable, and so it's part of the Lean, L-E-A-N acronym for the emergency medicines: **l**idocaine, **e**pinephrine, **a**tropine, and **n**aloxone.

O,

C, cocaine, works in the same way, but has abuse potential and is DEA Schedule II. Cocaine has some medical uses like eye drops.

K.

Next, we'll head to the cardio drugs, our second to last chapter.

CHAPTER 6: CARDIO MNEMONIC FLASHCARDS

Question 97. Name six drug classes we can use to treat hypertension.

ABCD

A_____, (RAAS drug acronym)

A_____, (RAAS drug acronym) and

A_____ B_____ (adrenergic antagonist)

B_____ B_____ (adrenergic antagonist)

C_____ C_____ B_____

D_____

97. Hypertension Treatments – ABCD

> **A**CEIs, ARBs, and alpha-1 blockers
> **B**eta-blockers
> **C**alcium channel blockers, CCBs
> **D**iuretics

Quick Summary

Hypertensive therapies depend on comorbidities. Unless the patient is African-American or has a contraindication, first-line treatment is an angiotensin-converting enzyme inhibitor (ACEI) or an angiotensin II receptor blocker (an ARB). After that, we use calcium channel blockers (CCBs) or diuretics. Finally, we employ beta-blockers or alpha-1 blockers. Learning the six drug classes is easy as ABCD.

A, ACEIs, angiotensin-converting enzyme inhibitors

A, ARBs, angiotensin II receptor blockers

A, alpha-1 blockers

B, beta-blockers

C, CCBs, calcium channel blockers

D, diuretics

Let's start with the renin-angiotensin-aldosterone system, RAAS.

Question 98. Identify eight points of the RAAS identifying where the four major RAAS drug classes work.

AR AA AR AR

A_____

R_____ - DRIs works here

A_____

A_____ C_____ E_____ ACEIs work here

A_____

R_____ F_____ A_____ ARBs work here

A_____ - AAs work here

R_____ - sodium and water

98. Renin Angiotensin Aldosterone System (RAAS) – AR AA AR AR

> **A**ngiotensinogen (the zymogen)
> **R**enin – **DRIs (direct renin inhibitors) work here**
>
> **A**ngiotensin I
> **A**ngiotensin converting enzyme (ACE) – **ACEIs here**
>
> **A**ngiotensin II
> **R**eceptor for Angiotensin II – **ARBs here**
>
> **A**ldosterone – **Aldosterone Antagonists, AAs, here**
> **R**etains renal sodium / water increasing blood pressure

Quick Summary

The renin-angiotensin-aldosterone system, RAAS, helps us regulate blood pressure, fluid, and electrolytes. Remember the eight cascade parts with: AR, AA, AR, AR.

A, angiotensinogen, is a zymogen, that

R, renin, an enzyme converts to

A, angiotensin I.

A, angiotensin-converting enzyme, ACE, converts angiotensin I to

A, angiotensin II, which binds to the

R, receptor for angiotensin II, to promote vasoconstriction and releases

A, aldosterone, to

R, retain sodium and water, further increasing blood pressure.

To lower blood pressure in a hypertensive patient, we block specific RAAS cascade parts.

Direct renin inhibitors, DRIs like **aliskiren (Tekturna)**, block the renin enzyme, so the cascade doesn't begin.

Angiotensin-converting enzyme inhibitors, ACEIs like **lisinopril**, block angiotensin II formation.

Angiotensin II receptor blockers, ARBs, like **losartan**, stop receptor binding.

Aldosterone antagonists, like **spironolactone**, block aldosterone formation blocking sodium and water retention.

Next, let's look at the specific medications in these categories.

CARDIO MNEMONIC FLASHCARDS

Question 99. Identify 7 ACEIs, the class suffix, and six side-effects, drug interactions, or important notes about the drug class.

ACE BLOCKER QUICK FACTS

B_____ (ACEI)

L_____ (ACEI)

O_____ H_____ (side effect)

C_____ (ACEI)

K+ _____ (side effect)

E_____ (ACEI)

R_____ (ACEI)

Qui_____ (ACEI)

C_____ (side effect)

K_____ (inflammatory mediator)

F_____ (ACEI)

A_____ (side effect)

C_____ P_____ (concern)

T

S_____ P_____ (concern)

Angiotensin-Converting Enzyme Inhibitors Suffix: -p_____

99. ACE INHIBITORS – ACE BLOCKER QUICK FACTS

Benazepril (Lotensin)

Lisinopril (Zestril)

Orthostatic hypotension

Captopril (Capoten)

K+ hyperkalemia

Enalapril (Vasotec)

Ramipril (Altace)

QUInapril (Accupril)

Cough

Kinin, bradykinin

Fosinopril (Monopril)

Angioedema

Contraindicated in pregnancy

T

Stenotic, post

Quick summary

ACE inhibitors block angiotensin I's change to angiotensin II reducing vasoconstriction and aldosterone secretion. ACE inhibitors are first-line hypertension agents with a mortality benefit in heart failure and MI patients. They are renal-protective for diabetics or patients with chronic kidney disease. There are many 'me-too' ACE inhibitors, so it's often easier to remember the 'pril, p-r-i-l" stem. Our mnemonic is ACE BLOCKER QUICK FACTS.

B, benazepril's brand **Lotensin** combines "lower" and "hypertension."

L, lisinopril, thrills an overworked heart, blocking angiotensin II from getting a start.

O, orthostatic hypotension is a rapid blood pressure decrease upon standing.

C, captopril's brand **Capoten** combines "captopril" and "hyper*ten*sion."

K+, hyperkalemia, results from increased potassium, putting patients at cardiac arrhythmia risk.

E, enalapril, is an oral form while **enalaprilat** ending in p-r-i-l-a-t is an injectable and the active enalapril metabolite. Brand **Vasotec** alludes to vasodilation and vasculature.

R, ramipril's brand **Altace** uses the "ace" from ACE inhibitor.

QUI, quinapril's brand **Accupril** also includes the –pril ending.

Memorizing Pharmacology Mnemonics

C, cough, usually dry, results from blocking bradykinin degradation and can happen any time after a patient starts an ACEI. We stop the drug to break a cough rather than use antitussives.

K, kinin, bradykinin, is the inflammatory mediator.

F, fosinopril's brand name **Monopril** has the ACE inhibitor ending –pril and "mono" for once daily dosing.

A, angioedema, is facial, lip, and mouth swelling that can interfere with breathing.

C, contraindicated, in pregnancy,

T,

S, stenotic, post. Watch for acute renal failure in the post-stenotic kidney or heart failure patients. ACEIs can cause acute renal injury because it dilates the kidney's efferent arterioles, increasing serum creatinine. However, we recommend ACEI for patients with chronic kidney disease and proteinuria.

Next, we'll look at an alternative ACEI mnemonic.

Question 100. Name six side effects to watch for with ACEIs.

CAPTOPRIL

C_____

A_____

P_____, H_____

T

O_____ H_____

P_____, D_____

R_____ F_____

I

L

Memorizing Pharmacology Mnemonics

100. ACEI SIDE EFFECTS - CAPTOPRIL

> **C**ough
> **A**ngioedema
> **P**otassium elevation as hyperkalemia
> **T**
> **O**rthostatic hypotension
> **P**regnancy, do not use
> **R**enal failure
> **I**
> **L**

Quick Summary

Another ACEI mnemonic that outlines the side effects but not drug names is CAPTOPRIL.

Now, we move on to ARBs.

CARDIO MNEMONIC FLASHCARDS

Question 101. Name five ARBS, the class suffix, and why you might switch a patient from an ACEI to an ARB.

VOCAL AID

V_____ (ARB)

O_____ (ARB)

C_____ (ARB)

A

L_____ (ARB)

A_____ C_____ (reason to switch)

I_____ (ARB)

D

Angiotensin II receptor blockers (ARBs) – suffix: –s_____

101. Angiotensin II Receptor Blockers (ARBs) – VOCAL AID

> **V**alsartan (Diovan)
> **O**lmesartan (Benicar)
> **C**andesartan (Atacand)
> **A**
> **L**osartan (Cozaar)
>
> **A**bsent cough
> **I**rbesartan (Avapro)
> **D**

Quick Summary

Angiotensin II receptor blockers, ARBs, block angiotensin II from binding to receptors preventing vasoconstriction and lowering blood pressure. ARBs, like ACE inhibitors, are first-line antihypertensive agents that slow renal disease and improve heart failure and heart attack mortality.

If a patient experiences angioedema with an ACE inhibitor, we often avoid an ARB. However, newer data suggests if the benefit outweighs the risk, such as in congestive heart failure, CHF, a patient might start on an ARB.

Learn angiotensin II receptor blockers by the suffix "sartan, s-a-r-t-a-n." Since ACEIs may cause a cough and ARBS don't because they don't interact with bradykinin, our mnemonic becomes VOCAL AID.

V, valsartan and brand **Diovan** share the letters v-a-n to help you connect them.

O, olmesartan's brand **Benicar** hints at benefit and cardiac.

C, candesartan, has a few letters from "decrease" blood pressure. Candesartan's brand **Atacand** takes seven of the generic name's letters.

A,

L, losartan 's brand **Cozaar** looks like it has R-A-A-S backward (for renin-angiotensin-aldosterone-system) with a "z" replacing the "s."

A, absent cough, highlights the ACEI and ARB difference.

I, irbesartan.

D.

Next, we cover alpha-1 blockers, another antihypertensive.

Memorizing Pharmacology Mnemonics

Question 102. Name three alpha-1 blockers, a class suffix, two therapeutic indications and a side effect.

BLOOD PRESSURE DIPPTS (DIPS)

D_____ (alpha-1 blocker)

I_____ vasodilation for BP (therapeutic use)

P_____ (alpha-1 blocker)

P_____, BPH (therapeutic use)

T_____ (alpha-1 blocker)

S_____, first dose (side effect)

Alpha-1 blocker class suffix: -a_____

Note: some alpha-1 blockers may not have this particular ending.

102. Alpha₁ Blockers for Hypertension – BLOOD PRESSURE DIPPTS (DIPS)

> **D**ox<u>azosin</u> (Cardura)
>
> **I**ncrease vasodilation for blood pressure
>
> **P**r<u>azosin</u> (Minipress)
>
> **P**rostate, Benign prostatic hyperplasia, BPH
>
> **T**er<u>azosin</u> (Hytrin)
>
> **S**yncope, first-dose

Quick Summary

As a treatment for **benign prostatic hyperplasia, BPH,** alpha blockers relax the bladder neck and smooth muscle, to improve urine flow. We'll look at more specific drugs for this indication in Chapter 7.

In **hypertension,** alpha-1 receptor antagonists block alpha-1 receptors, causing peripheral vessel vasodilation reducing blood pressure. With better antihypertensives available, we only use them in resistant hypertension.

The dizziness side effect often manifests as first-dose syncope, a sudden drop in blood pressure on a patient's first dose. Give at bedtime to cancel this orthostatic effect. Our mnemonic is blood pressure, BP DIPPTS, D-I-P-P-T-S.

D, dox<u>azosin</u>, has the "azosin, a-z-o-s-i-n" alpha-blocker stem. Brand **Cardura** provides <u>dura</u>ble <u>car</u>diac relief using the d-u-r-a from durable and c-a-r from cardiac to form **Cardura**.

I, increase vasodilation, the drug class' antihypertensive mechanism of action.

P, prazosin,'s brand name **Minipress** alludes to a mini or smaller blood pressure.

P, prostate, reminds us of the benign prostatic hyperplasia (BPH), alpha-blocker therapeutic indication.

T, terazosin also has prazosin's –azosin stem. Brand **Hytrin** takes six letters from "hypertension."

S, syncope, first-dose, is a sudden blood pressure drop from the first dose. Give at bedtime to avoid the issue.

Our next group, beta-blockers, can reduce heart rate to help lower blood pressure.

Question 103. Name a first-generation beta-blocker, the class suffix, why it is a good choice for migraine prophylaxis, a drug-disease contraindication and the beta receptors it affects. Name three second-generation beta-blockers, the beta receptor they affect, and the organ that beta-receptor affects. Name two non-selective third generation beta blockers, their two class suffixes, and the three receptors they affect. Name a third-generation beta-blocker, its class suffix, and whether it is selective or not, and a possible side effect of that specific drug based on a unique mechanism.

PLAN FOR EXAMS, GET COLA NOW

P_____ (1st-generation beta-blocker)
L_____ (means fat-loving)
A_____ (contraindication)
N_____ for beta-1 and beta-2 receptors

E_____ (2nd-generation beta-blocker)
X
A_____ (2nd-generation beta-blocker)
M_____ (2nd-generation beta-blocker)
S_____ for beta-1 receptors

C_____ (3rd generation beta-blocker)
O
L_____ (3rd generation beta-blocker)
A_____, B_____ and B_____

N_____ (3rd-generation beta blocker)
O_____ Beta-1
W_____ (concern)

103. BETA BLOCKERS BY GENERATION – PLAN FOR EXAMS, GET COLA NOW

Propran<u>olol</u> (Inderal LA) (1ST)

Lipophilicity

Asthma contraindication

Non-selective for Beta receptors (Beta-1, Beta-2)

Esm<u>olol</u> (Brevibloc) (2ND)

X

Aten<u>olol</u> (Tenormin) (2ND)

Metopr<u>olol</u> (Toprol XL, Lopressor) (2ND)

Selective for Beta-1 receptors in the heart

Carve<u>dil</u>ol (Coreg) (3RD)

O

Labet<u>alol</u> (Normodyne) (3RD)

Alpha-1, Beta-1, and Beta-2

Nebiv<u>olol</u> (Bystolic) (3RD)

Only Beta-1

Watch for hypotension with nitrous oxide mechanism

Quick summary

While we don't often use beta blockers for hypertension, we use them for heart failure, heart disease, and post-MI. We find mostly beta-1 receptors in the heart and beta-2 receptors in the lungs. You have *one* heart and *two* lungs, which makes this easy to remember. Watch for bradycardia, fatigue, dizziness, and masking of hypoglycemia signs. Our mnemonic to sort beta-blockers by generation is PLAN FOR EXAMS, GET COLA NOW, as exams and caffeine both impact heart rate.

Mnemonic 1. PLAN

P, propranolol (1ST generation), is a non-selective beta blocker blocking both beta-1 and beta-2 receptors. Blocking beta-1 receptors reduces heart rate. Blocking beta-2 inhibits bronchodilation, which may cause asthmatics problems.

The "o-l-o-l" beta-blocker stem looks like two-letter "b's" backward to help you remember (b) beta (b) blocker. Alternatively, if you "oh, laugh out loud," using the o-l-o-l. Imagine heart rate goes down with a good laugh. Brand **Inderal** blocks "all" beta-receptors, both one and two.

L, lipophilicity, of propranolol, allows it to cross the blood-brain barrier for migraine prophylaxis and anxiety.

A, asthma contraindication, with beta-2 receptor blockade.

N, non-selective for Beta receptors, (Beta-1, Beta-2),

Mnemonic 2. EXAMS

E, esmolol (2ND generation),

X,

A, atenolol (2ND generation) is beta-1 selective, only affecting the heart. The "Ten, T-e-n" in Brand **Tenormin** connects the "ten, t-e-n" in **atenolol**.

M, metoprolol (2ND generation), has two salt forms – twice daily tartrate, and once daily succinate. Metoprolol tartrate's brand **Lopressor** lowers blood pressure, using the l-o from lowers and p-r-e-s-s from pressure – Lopressor. **Metoprolol succinate**, the long-acting form is **Toprol XL**. The X-L, used to identify extra-large clothing, serves to indicate an extra-long acting effect.

S, selective for Beta-1 receptors.

Mnemonic 3. COLA.

C, carvedilol (3RD generation), is a non-selective beta-blocker, that additionally blocks alpha-1 receptors to vasodilate vessels. Carvedilol's hybrid stem has "dil, d-i-l" replacing the first "o" in "olol, o-l-o-l," to combine vasodilation and beta-blockade. Carvedilol has approval for heart failure to reduce mortality. Brand **Coreg** regulates coronary function, using the r-e-g from regulates and the c-o from coronary – **Coreg**.

O,

L, labetalol (3RD generation), is another non-selective beta-blocker with vasodilator effects. Labetalol may be a good option for hypertensive pregnant women. Labetalol has "beta" from beta-blocker in the name. Instead of "–olol" for beta-blocker, the stem is "–alol" for alpha / beta-blocker.

A, alpha-1, Beta-1, and Beta-2 blockade.

Mnemonic 4. NOW.

N, nebivolol (3RD generation), is beta-1 selective that causes vasodilation through nitric oxide, which can lower blood pressure. Brand **Bystolic** takes letters from sy<u>stolic</u> (the top blood pressure number) and dia<u>stolic</u> (the bottom number).

O, only Beta-1,

W, watch for hypotension.

I've also seen many mistaken beta-blocker generation mnemonics, and we'll look at that next.

Question 104. Identify an incorrect mnemonic that errs as a cognitive heuristic bias.

DO NOT USE A through N, O through Z

A_____ (selective beta-blocker)

M_____ (selective beta-blocker)

N_____ *(non-selective beta-blocker)*

make A through N are selective beta-blockers untrue.

P_____ *(non-selective beta-blocker)*

C_____ (selective beta-blocker) &

C_____ (selective beta-blocker)

make O through Z are non-selective beta-blockers untrue.

104. Beta Blockers' Selectivity – Do Not Use: A Through N, O Through Z

> ### A through N (Selective)
>
> **A**ten<u>olol</u> (Tenormin)
>
> **M**etopr<u>olol</u> (Toprol XL, Lopressor)
>
> **N**ad<u>olol</u> (Corgard) makes this untrue.
>
> ### O through Z (Non-selective)
>
> **P**ropran<u>olol</u> (Inderal)
>
> **C**arte<u>olol</u> (Cartrol) & Carve<u>dilol</u> (Coreg) make this untrue.

Quick Summary

If a student has never seen nadolol, carteolol, and carvedilol, they might wrongly believe all A through N beta-blockers are selective. They might also believe all O through Z are non-selective. These incorrect shortcuts are cognitive heuristic biases. In plain English, this all means the student doesn't know what he doesn't know and makes a mistake. In health care, it's much worse to make a statement that you think is correct rather than say, "I don't know." Next, we'll move to beta-blocker concerns to reinforce some of the side effects.

Question 105. Name five concerns when using a beta blocker.

ABCDE

A_____ with non-selective propranolol

B_____ as heart block

C_____

D_____ M_____ signs of low blood sugar

E_____ A_____, hyperkalemia

105. Beta-blocker concerns - ABCDE

> **A**sthma, nonselective such as propran<u>olol</u>
>
> **B**lock as heart block
>
> **C**OPD
>
> **D**iabetes mellitus, masked signs of low blood sugar
>
> **E**lectrolyte abnormalities, hyperkalemia

Quick Summary

The ABCDE mnemonic serves as an anchor for these five concerns.

Another antihypertensive class is the calcium channel blocker, CCB, which we'll take a look at next.

Question 106. Describe a disease state non-dihydropyridine calcium channel blockers can treat that dihydropyridines can't, name two dihydropyridine CCBs and their suffixes.

SLOW DVD

D_____ (disease state)

V_____ - class suffix -p_____

D_____ - class suffix -t_____

106. Calcium Channel Blockers (CCBs) – Non-Dihydropyridines – SLOW DVD

> **D**ysrhythmias
> **V**era<u>pamil</u> (Calan)
> **D**il<u>tiazem</u> (Cardizem)

Quick Summary

We divide calcium channel blockers, CCBs, into two classes – dihydropyridines and non-dihydropyridines. Both vasodilate, but only the non-dihydropyridines block the AV and SA node to affect dysrhythmias.

V, vera<u>pamil</u>, and D, dil<u>tiazem</u>, are both non-dihydropyridine-type CCBs that stop calcium ions from entering smooth muscle and the myocardium. They can decrease heart rate and slow AV node conduction. You may hear negative chronotropy, slowing heart rate, or negative inotropy, slowing ventricular contraction. While we see non-dihydropyridine CCBs for dysrhythmias like atrial fibrillation, they work for hypertension and angina also. Side effects include bradycardia, AV block, hypotension, and constipation, which is more likely with vera<u>pamil</u>. Other side effects include edema, which can worsen symptoms of heart failure, and dizziness.

We use the mnemonic SLOW DVD because these drugs can slow heart rate and ventricular contraction.

D, dysrhythmias, are abnormal heart rhythms these drugs can treat.

V, verapamil has the "pamil, p-a-m-i-l" stem. One student came up with "Vera and Pam are ill and need this calcium blocking cardiac pill," adding the V-e-r-a from Vera and P-a-m from Pam to make verapamil. Often, we associate verapamil with constipation. My grandmother, a Navy nurse, used to put a verapamil tablet on my grandfather's breakfast cereal spoon. I always thought my grandfather was silently praying before he ate. When I asked him why he was so quiet, he said something to the effect of, "I'm deciding whether I want to eat or poop today." The brand name **Calan** takes c-a-l from the word calcium, and a-n from channel blocker, to make C-a-l-a-n, Calan.

D, diltiazem, has the "–tiazem, t-i-a-z-e-m" stem. Brand **Cardizem** adds the first five letters from cardiac, c-a-r-d-i to the last three letters of the generic diltiazem, z-e-m, to make Cardizem.

In the next section, we'll continue with dihydropyridine CCBs and another mnemonic.

CARDIO MNEMONIC FLASHCARDS

Question 107. Name five dihydropyridine calcium channel blockers, their class suffix, and a side-effect to watch out for.

SAVED INFANTS

I_____ (CCB)

N_____ (CCB)

F_____ (CCB)

A_____ (CCB)

N_____ (CCB)

T_____, rebound (side effect)

Dihydropyridine calcium channel blockers suffix –d_____

107. Calcium Channel Blockers (CCBs) – Dihydropyridines – SAVED INFANTS

> **I**sra<u>dipine</u> (Dynacirc)
>
> **N**ife<u>dipine</u> (Procardia)
>
> **F**elo<u>dipine</u> (Plendil)
>
> **A**mlo<u>dipine</u> (Norvasc)
>
> **N**icar<u>dipine</u> (Cardene)
>
> **T**achycardia, rebound

Quick summary

The SAVED INFANTS mnemonic is a personal one. My wife, pregnant with triplets, faced labor at 19 weeks as one daughter tried to get out. Magnesium sulfate stopped the contractions and allowed the surgeon to perform a cerclage procedure. However, we needed to avoid further contractions. The doctor prescribed nifedipine, a dihydropyridine calcium channel blocker.

As with non-dihydropyridines, blocking calcium channels produces arterial vasodilation, decreasing vascular resistance and blood pressure and anginal pain. Dihydropyridines don't affect impulse conduction through the AV node for dysrhythmias.

CCBs are first-line agents for hypertensive African-American patients. As with other vasodilating CCBs, side effects include dizziness and hypotension. They can cause edema, but less than non-dihydropyridines making them the better choice in heart failure patients who must use a CCB. All dihydropyridines have the "dipine, d-i-p-i-n-e" stem taking d-i-p-i-n-e from dihydropyridine. Also notice the dip, d-i-p, in, i-n blood pressure. Let's visit the SAVED INFANTS mnemonic.

I, isradipine,

N, nifedipine's brand **Procardia** takes the "p-r-o" from "promotes" and "c-a-r-d-i-a" from "cardiac." **Procardia** promotes cardiac health.

F, felodipine,

A, amlodipine's brand **Norvasc** includes "n-o-r" from normalizes and "v-a-s-c" from vasculature – **Norvasc**.

N, nicardipine,

T, tachycardia, rebound. Dihydropyridine CCBs, with peripheral vasodilation and no cardiac suppression, can cause rebound tachycardia. Our body tries to compensate for the low blood pressure. Closely watch short-acting nifedipine during first administration, as profound hypotension can occur.

Next, we'll turn to diuretics.

Memorizing Pharmacology Mnemonics

Question 108. Name five diuretics in at least four unique classes traveling from the glomerulus to PCT, Loop of Henle, DCT, and collecting duct and two of their class suffixes. Identify two diuretics that cause hypokalemia and two that cause hyperkalemia.

MAN, FLUSH THIS

Man_____ (PCT, osmotic diuretic)

F_____ (Loop) - class suffix -s_____

L_____ diuretic

U

S

H_____ (DCT) - class suffix -t_____

T_____ (Collecting duct, potassium sparing)

H_____ (Thiazide diuretic)

I__ C_____ for hypokalemia (Dyazide, Maxzide)

S_____ (Collecting duct)

366

108. Diuretics – MAN, FLUSH THIS

> **M**ANnitol (Osmitrol) (PCT)
>
> **F**urosemide (Lasix) (Loop)
> **L**oop diuretic
> **U**
> **S**
> **H**ydrochlorothiazide (Microzide, Hydrodiuril) (DCT)
>
> **T**riamterene (Dyrenium) (Collecting duct)
> **H**ydrochlorothiazide
> **I**n combination for hypokalemia (Dyazide, Maxzide)
> **S**pironolactone (Aldactone) (Collecting duct)

QUICK SUMMARY

The nephron, or smallest functional kidney unit, winds and twists like a slide from glomerulus to collecting duct. The diuretic's diuresis coincides with proximity to the glomerulus. Since diuretics make you pee, our mnemonic is MAN, FLUSH THIS.

MAN, mannitol, increases osmotic pressure, inhibiting water reabsorption and electrolytes causing urinary excretion. Mannitol is for stroke patients and other brain injuries to reduce intracranial pressure. Brand **Osmitrol** combines "osmotic" and control brain swelling, taking the o-s-m-i from osmotic and t-r-o-l from control – **Osmitrol**. The actor Bruce Lee died from an intracranial event.

F, furosemide, is an **L, loop diuretic**, that works at the ascending loop of Henle. Loop diuretics block sodium and chloride reabsorption. Water follows sodium, so if sodium remains in the loop, water follows to urination. We see furosemide more in heart failure and volume overload than hypertension.

The "semide, s-e-m-i-d-e" stem indicates a "furosemide-type" diuretic like its cousin torsemide. Once student invented "I have to pee furiously" as her mnemonic since loop diuretics produce significant diuresis, taking the f-u-and r from furosemide. Brand **Lasix** indicates it lasts six hours; **Lasix**.

We try to dose diuretics in the morning to avoid nighttime urination. Side effects include ototoxicity, dizziness, hyponatremia, hypokalemia, and hypocalcemia. We often prevent hypokalemia with a potassium chloride supplement.

U, S,

H, hydrochlorothiazide, or HCTZ, is a thiazide diuretic. Like furosemide, HCTZ inhibits sodium reabsorption, but at the distal convoluted tubule. HCTZ causes sodium, chloride, and water excretion. Side effects include low sodium and potassium, but HCTZ differs from loop diuretics in that it produces increased calcium and uric acid levels. Other side effects include dizziness and phototoxicity.

Thiazide diuretics end in "–thiazide, t-h-i-a-z-i-d-e." Hydrochlorothiazide's abbreviation H-C-T-Z comes from "h" for <u>h</u>ydro, "c" for <u>c</u>hloro, "t" for <u>t</u>hia, and "z" for <u>z</u>ide. Thiazides don't produce as much diuresis as loop diuretics, but are good first-line antihypertensives. While the "hydro" in **hydrochlorothiazide** stands for the <u>hydro</u>gen atom, you can think of "hydro" as "water" for diuretic. Brand **Microzide** has thia<u>zide</u>'s last four letters, z-i-d-e. Brand **Hydrodiuril** combines "<u>hydro</u>" for water and "<u>diur</u>esis."

T, triamterene, is a potassium-sparing diuretic, and

H, hydrochloro<u>thiazide</u>, a thiazide diuretic, works

I, in combination for hypokalemia, in a single pill as brand **Dyazide** capsules or brand **Maxzide** tablets.

S, spironolactone, is potassium-sparing, but not technically a diuretic. This aldosterone antagonist increases sodium and water excretion in the collecting duct. Spironolactone conserves potassium, so *hyper*kalemia is a concern. Spironolactone and other potassium-sparing diuretics aren't first-line antihypertensives. They treat resistant hypertension and later heart failure stages. Other spironolactone side effects include female irregular menses and male gynecomastia.

To remember **spironolactone** works in the collecting duct, the last nephron structure, I look at the "lactone" and think "last one," taking the "l-a" and "o-n-e" from lactone. The brand name replaces the "spironol" of **spironolactone** with "ald, a-l-d" to make **Aldactone**. The "ald" is crucial because **spironolactone** blocks <u>ald</u>osterone, a steroid hormone that guides sodium and water retention.

Let's look at an alternate mnemonic for the diuretic classes in our next slide.

Memorizing Pharmacology Mnemonics

Question 109. Name five diuretic classes and at least three drug class suffixes.

LOCATE A BATHROOM SOON

L_____ - class suffix -s_____

O_____

C_____ - class suffix -z_____

A_____ I_____

T_____ - class suffix -t_____

E

109. Diuretic classes – LOCATE A BATHROOM SOON

> *L*oop, furo<u>semide</u> (Lasix)
>
> *O*smotic, mannitol (Osmitrol)
>
> *C*arbonic anhydrase inhibitor, dor<u>zolamide</u> (Trusopt)
>
> *A*ldosterone inhibitor, spironolactone (Aldactone)
>
> *T*hiazide, hydrochloro<u>thiazide</u>, (Hydrodiuril)
>
> *E*

Quick Summary

This other mnemonic, LOCATE A BATHROOM SOON, prioritizes the diuretic classes, then the drug names.

Next, we'll tackle hydrochlorothiazide, HCTZ, indications.

Question 110. Name four indications for hydrochlorothiazide.

HCTZ

H_____

C_____ H_____ F_____ (CHF)

T_____, (diabetes insipidus is counterintuitive)

Z's_____, kidney stones

110. Hydrochlorothiazide indications – HCTZ

> *H*ypertension
>
> *C*ongestive heart failure (CHF)
>
> *T*hirst, e.g., diabetes insipidus is counterintuitive
>
> *Z'*stones, kidney stones

Quick summary

While these indications are self-explanatory, HCTZ's only role in CHF is for sequential nephron blockade. They use it 30 minutes before a loop, not as a solo diuretic.

Next, we'll look at how diuretics cause electrolyte imbalances.

Memorizing Pharmacology Mnemonics

Question 111. Name four drug classes that cause hyperkalemia and at least one example drug from each class.

POTASSIUM PANDA

P_____ sparing diuretics - drug _____

A_____ inhibitors - drug _____

N_____ - drug _____

D

A_____ - drug _____

111. Drug classes that increase potassium – POTASSIUM PANDA

> **P**otassium-sparing diuretics (spironolactone, triamterene)
>
> **A**CEI (lisino<u>pril</u>, enala<u>pril</u>)
>
> **N**SAIDs (aspirin, ibu<u>profen</u>, naproxen)
>
> **D**
>
> **A**RBs (val<u>sartan</u>, lo<u>sartan</u>)

Quick summary

A PANDA is big, so a POTASSIUM PANDA is "big potassium" or hyperkalemia. Check potassium in patients on these medications.

Next, we go into detail about cholesterol lowering statins.

Memorizing Pharmacology Mnemonics

Question 112. Name six HMG Co-As, the class suffix, the primary lipid target, and a disconcerting sign or symptom.

LDL FALLS SHARPLY

F_____ (HMG Co-A)

A_____ (sign/symptom)

L_____ (fat-loving)

L_____ (primary lipid target)

S

S_____ (HMG Co-A)

H_____ intensity statins:

A_____ (high intensity HMG Co-A)

R_____ (high intensity HMG Co-A)

P_____ (HMG Co-A)

L_____ (HMG Co-A)

Y

HMG-CoA Reductase Inhibitor suffix: -s_____

376

112. Cholesterol lowering – HMG Co-A reductase inhibitors - STATINS – LDL FALLS SHARPLY

> *F*luva<u>statin</u> (Lescol)
> *A*ches, muscle
> *L*ipophilic
> *L*DL
> *S*
>
> *S*imva<u>statin</u> (Zocor)
> *H*igh-intensity statins atorvastatin and rosuvastatin
> *A*torva<u>statin</u> (Lipitor)
> *R*osuva<u>statin</u> (Crestor)
> *P*rava<u>statin</u> (Pravachol)
> *L*ova<u>statin</u> (Mevacor, Altoprev)
> *Y*

Quick Summary

HMG-CoA reductase is an enzyme that converts HMG-CoA to mevalonate, the rate-limiting step in cholesterol synthesis. Statins are the drug of choice for lowering LDL, and also help increase HDL and triglycerides. Current guidelines recommend statins for patients with clinical <u>A</u>thero<u>S</u>clerotic <u>C</u>ardio<u>V</u>ascular

Disease, ASCVD. Example conditions include: stroke, transient ischemic attacks, coronary artery disease, patients with elevated LDL, specific diabetic patients, and people with > 5%, 10-year-risk of ASCVD. Patients tolerate statins well, except for possible myopathy or muscle pain which may lead to rhabdomyolysis, a condition of muscle breakdown and kidney failure risk. Statins can raise liver enzymes.

Currently, prescribers determine statin doses based on patient condition groupings for primary or secondary ASCVD prevention. Statins have three dosing categories – low, moderate, and high intensity. Many statins are 'me-too' drugs, but there are some subtle differences. I've put these three groups together in the mnemonic.

Atorvastatin and **rosuvastatin** serve as **high and moderate intensity.**

Lovastatin, simvastatin, and **atorvastatin** are **lipophilic** and may go into muscle more readily.

Rosuvastatin and **pravastatin** might solve the muscle pain problem, as they are **hydrophilic**. However, there isn't a lot of data on this.

Statins, or HMG-CoA reductase inhibitors, help with hyperlipidemia by lowering LDL, the bad cholesterol. The mnemonic for statins is LDL FALLS SHARPLY.

Mnemonic 1. FALLS.

F, fluvastatin, has the substem "va, v-a" and adds s-t-a-t-i-n to get "vastatin, v-a-s-t-a-t-i-n" as a way to identify the H-M-G-Co-As in general.

A, aches, muscle, like myopathy possibly ending as rhabdomyolysis.

L, lipophilic, or fat loving.

L, LDL, for low density-lipoprotein, the bad cholesterol.

S.

Mnemonic 2. SHARPLY

S, simvastatin,

H, high intensity statins atorvastatin and rosuvastatin,

A, atorvastatin's brand **Lipitor**, is a lipid gladiator, taking l-i-p from lipid and t-o-r from gladiator – Lipitor.

R, rosuvastatin's brand **Crestor** decreases cholesterol, taking the c-r from decreases and e-s-t-o-r from cholesterol. Crestor.

P, pravastatin, has the "chol" in pravastatin's **Pravachol**, which hints at cholesterol.

L, lovastatin, has "cor, c-o-r" in lovastatin's **Mevacor**, which hints at coronary.

Y.

Next, we'll look closely at HMG Co-A side effects.

Question 113. Name two potential side effects of HMG Co-As, whether you should use them in pregnancy, the drug's primary lipid target, and the enzyme it affects to exert its influence.

HMG Co-A

H_____ (side effect)

M_____ and rhabdomyolysis (side effect)

G_____, don't use in pregnancy

C_____. LDL lowering effect

O_____ of depression is controversial

A_____ the enzyme HMG Co-A

113. HMG-CoA REDUCTASE INHIBITOR (STATIN) DETAILS – HMG Co-A

> **H**epatotoxicity
>
> **M**yopathy to rhabdomyolysis
>
> **G**estation, don't use in pregnancy
>
> **C**holesterol LDL lowering therapeutic effect
>
> **O**nset of depression is controversial
>
> **A**ffects an enzyme, HMG Co-A

QUICK SUMMARY

Students use letters of the H-M-G-Co-A class to memorize primary potential adverse effects and therapeutic and mechanistic details.

Next, we tackle a newer lipid-lowering group, the PCSK9s.

Question 114. Name two PCSK9 inhibitors and their role in cholesterol lowering.

THE PCSK9 PEACH

P_____ inhibitors

E_____ (PCSK9 inhibitor)

A_____ (PCSK9 inhibitor)

C_____ 2nd line agents

H_____ high cholesterol

114. PCSK9 Inhibitors – THE PCSK9 PEACH

> **PCSK9** inhibitors
> **E**volocu<u>mab</u> (Repatha)
> **A**lirocu<u>mab</u> (Praluent)
> **C**holesterol 2nd line agents
> **H**ereditary high cholesterol

Quick summary

The PCSK9 inhibitors are second-line treatments for elevated cholesterol that didn't respond to diets or statins like atorvastatin. Our mnemonic is the PCSK9 PEACH because peaches are healthy as is lower cholesterol. Just adding a few words to the mnemonic allows us to make an extended sentence.

P, PCSK9 inhibitors (proprotein convertase subtilisin/kexin type 9) are:

E, evolocu<u>mab</u> and

A, alirocu<u>mab</u> which we use as

C, cholesterol 2nd-line agents for

H, hereditary high cholesterol.

Next, we look at other lipid-lowering medications that are second-line to statins.

Memorizing Pharmacology Mnemonics

Question 115. Name six non-statin lipid lowering agents, the unofficial class infix for fibrates, the fibrates' mechanism of action and target lipid molecule.

FIGHT ONCE AGAIN

F_____ (fibrate)

I

G_____ (fibrate)

H_____

T_____ (target lipid molecule)

O_____-3-acid ethyl esters (fish oil derivative)

N_____ (Vitamin B3)

C_____ (Bile acid sequestrant)

E_____ (Absorption inhibitor)

Unofficial class infix for fibrates: _-f_____

115. Cholesterol-Lowering Agents – Fight Once Again

Fenofibrate (Tricor)
I
Gemfibrozil (Lopid)
High
Triglycerides

Omega-3-acid ethyl esters (Lovaza)
Niacin (Niaspan)
Colesevelam (Welchol)
Ezetimibe (Zetia)

Quick Summary

With very high triglycerides, LDL may increase. Fibrates can increase HDL, good cholesterol. We see their use in patients with triglyceride levels over 500. Similar to what we see with statins, myopathies and liver enzyme increases may happen. Other side effects include nausea and increased gallstone risk. The mnemonic FIGHT ONCE AGAIN reminds us that even with statin therapy, we may need to continue to add agents especially if triglycerides are high.

F, **fenofibrate**, increases VLDL breakdown, decreasing triglyceride levels. The brand **TRICOR** takes the t-r-i from triglycerides and c-o-r from coronary – TRICOR.

I,

G, **gemfibrozil,** is a fibrate similar to fenofibrate, and a drug of choice for elevated triglycerides. Gemfibrozil with statins increases rhabdomyolysis risk. You can see the "fib" in the fibric acid derivative, gemfibrozil. Brand **Lopid** hints at "lowering lipids."

H, **high,** and T, **triglycerides**, form the target of those two drugs.

O, **omega-3 acid ethyl esters**, or fish oil or omega-3 fatty acids, have an unknown mechanism of action. We use omega-3 fatty acids for hypertriglyceridemia, but not as monotherapy. Like bile acid sequestrants below, fish oil has a high pill burden and a bad taste. While OTC products have omega-3 fatty acids, the FDA does not regulate these supplements. Only prescription medications like **Lovaza** have tested fish oil quantities.

N, **niacin**, nicotinic acid or vitamin B3, reduce VLDL and LDL production, which leads to triglyceride and LDL reduction. Niacin also decreases HDL clearance. Patients must take niacin at high doses to get good cholesterol-lowering effects, and immediate release formulations lead to flushing, itching, and dry skin. Non-flushing formulations exist, but don't seem to lower lipid levels. Patients can improve tolerability by taking niacin in the evening with a low-fat snack, taking aspirin before niacin, or taking the extended-release formula. Other side effects include hyperuricemia and nausea. It may also cause facial flushing that an **aspirin** thirty minutes before treatment prevents. This flushing, unfortunately, gets in the way of helping many patients who would otherwise benefit from an inexpensive cholesterol-lowering medication.

C, colesevelam, a bile acid sequestrant, binds intestinal bile acids, which then leave through feces. They can prevent bile acid reabsorption through enterohepatic circulation. Bile acid sequestrants are mild LDL lowering agents with little effect on triglycerides. They have a high pill burden with multiple tablets or packets taken many times daily. Because they bind other substances in the GI tract, they often impair vitamins and mineral absorption and other medications as well. Colesevelam has fewer drug interactions than the other bile acid sequestrants. Common side effects include constipation, abdominal pain, and bloating. A small benefit comes from binding glucose in people with type-2 diabetes for a small A1C reduction. **Colesevelam** has both a generic and a brand name with a hint at <u>chol</u>esterol and the brand name at getting "well," **Welchol**.

E, eze<u>ti</u>mibe, inhibits small intestine cholesterol absorption. It decreases LDL with no effect on HDL or triglycerides. Ezetimibe use, statin use, risks increased liver enzymes and myopathy. Don't use as a monotherapy – instead, add it to a statin to further lower LDL. It can combine with an HMG-CoA like simvastatin (Zocor) to make Vytorin.

Now, let's take a look at anticoagulants.

Memorizing Pharmacology Mnemonics

Question 116. Name four anticoagulants and two class suffixes.

HAS FEWER CLOTS

H_____ -class suffix -p_____

A_____ protamine sulfate

S_____ half-life

F_____ - class infix -p_____

E_____ - class suffix -p_____

W_____ - class suffix -f_____

E

R

388

116. Anticoagulants – HAS FEWER CLOTS

> **H**eparin (UFH)
> **A**ntidote protamine sulfate
> **S**hort half-life
>
> **F**ondaparinux (Arixtra)
> **E**noxaparin (Lovenox)
> **W**arfarin (Coumadin)
> **E**
> **R**

Quick Summary

Anticoagulants work on the clotting cascade to prevent new clots from forming and existing clots from getting bigger. They do not break down formed clots. Prescribers use them for stroke and venous thromboembolisms (VTEs) like pulmonary embolism (PE) and deep vein thrombosis (DVT). As such, the mnemonic is HAS FEWER CLOTS.

H, heparin, is an injectable that binds to antithrombin and inhibits thrombin preventing fibrinogen conversion to fibrin, to form a stable clot. Heparin infuses continuously with existing clots or is post-MI. Heparin as a subcutaneous injection helps prevent hospitalized patients' clots. Heparin can keep IV lines open with a flush. Heparin can cause heparin-induced thrombocytopenia, or HIT, where antibodies attack heparin and cause blood clots. Heparin and "bleedin'" sort of rhyme to help us remember its primary adverse effect. A student of mine mentioned the actor Dennis Quaid's twins, who received a double dose of heparin that caused bleeding. Sometimes, knowing a celebrity who has dealt with a condition helps place the drug in memory.

A, antidote is protamine sulfate for immediate heparin reversal.

S, short half-life. The liver metabolizes heparin, and it has a short half-life, so it's a good choice for renal impairment and those going to surgery.

F, fondaparinux, is an indirect factor Xa inhibitor. It is a subcutaneous injection, and for the treatment and prevention of VTE. There are no current antidotes for fondaparinux. We see this for patients with a history of HIT because it's entirely synthetic.

E, enoxaparin, is a low-molecular-weight heparin that works like heparin in binding to antithrombin. It, however, is more specific for factor Xa than factor IIa. Like heparin, enoxaparin prevents blood clots and heart attacks. It is always a subcutaneous injection. The kidneys clear enoxaparin and may require dose reduction. It also has a longer half-life, making it a poor choice for emergency procedures. Enoxaparin has less HIT risk than heparin and we prefer it for cancer patients. The reversal agent is also protamine. Recent studies show Xa inhibitors to be non-inferior. Enoxaparin and heparin share the "parin, p-a-r-i-n" stem because they are related. It is also more expensive per dose, but patients can use it at home. It's also for bridge therapy for patients starting **warfarin**. Brand **Lovenox** is low-molecular-weight heparin for deep vein thrombosis prevention, taking the l-o from low and v-e-n from vein, **Lovenox**.

W, **warfarin**, is a vitamin K reductase inhibitor that causes depletion of factors II, VII, IX, and X, as well as proteins C and S. It is for VTE, atrial fibrillation, and clot prevention in those with artificial valves. Warfarin's dosing is patient specific and based on the International Normalized Ratio, or INR. The goal INR will vary on the indication but is typically around 2-3. If the INR is below 2, that means the blood is not adequately anticoagulated, and the patient is at risk of a clot. If the INR is above 3, that means the blood could potentially be too anticoagulated, thus putting the patient at risk of bleeds. Many things can affect a patient's INR. Missed doses and increased foods or supplements with vitamin K can decrease the INR. Increased alcohol intake and acute illness can contribute to increases in INR. Warfarin has many CYP P450 interactions, so various medications can increase or decrease the INR. A significant warfarin side effect is bleeding. Patients should monitor for bruising, nose or gum bleeding, and bloody stool. The reversal agent for warfarin is vitamin K, or phytonadione.

Note that the "parin, p-a-r-i-n" stem from the anticoagulants **heparin** and **enoxaparin** and "farin, f-a-r-i-n" stem of **warfarin** are similar. This reminds students they are all anticoagulants. Students associate bleeding with warfare and warfarin has the w-a-r-f-a-r from warfare to help us recognize bleeding as a potential adverse effect. The I-N-R (international normalized ratio), a way of measuring **warfarin's** effectiveness, monitors the patient who is on therapy. "I-N-R" happen to be the last three letters of **warfarin** to help you remember this, as well. A student of mine said that **warfarin** has "far, f-a-r" in it, as in you have to go far to have blood drawn. A way to remember that Vitamin K affects warfarin's brand name **Coumadin** and coagulation is to spell **Coumadin** with a "K" instead of a "C." K-o-u-m-a-d-i-n.

E,

R.

Next, we visit warfarin, which has lots of interactions.

CARDIO MNEMONIC FLASHCARDS

Question 117. Name eleven warfarin-drug medication interactions.

ACADEMIC'S FAB-4

A

C_____ (H2 blocker)

A_____ (NSAID)

D_____ (Antibiotic for leprosy)

E_____ (Macrolide antibiotic)

M

I_____ (NSAID)

C_____ (Fluoroquinolone)

S_____ (HMG Co-A reductase inhibitor)

F_____ (Triazole antifungal)

A_____ (Antiarrhythmic)

B_____ (Sulfa antibiotic brand name)

4 (M_____) (Nitroimidazole antibiotic)

117. WARFARIN: INTERACTIONS – ACADEMIC'S FAB-4

> **A**
> **C**imet<u>id</u>ine (Tagamet)
> **A**spirin (Ecotrin)
> **D**apsone
> **E**ry<u>thromycin</u> (E-mycin)
> **M**
> **I**ndo<u>methac</u>in (Indocin)
> **C**ipro<u>floxac</u>in (Cipro)
> **S**im<u>vas</u>tatin (Zocor)
>
> **F**lu<u>cona</u>zole (Diflucan)
> **A**miodarone (Cordarone)
> **(B**actrim), generic is <u>sulfa</u>methoxazole / trimetho<u>prim</u>
> **4** Met<u>roni</u>dazole (Flagyl)

QUICK SUMMARY

Our biggest concern with warfarin is the increased INR or bleed. When starting one of these medications while on warfarin, we may have to reduce a warfarin dose. Note there are the "FAB-4" because they have possible life-threatening interactions with many elderly patient indications. Next, we visit direct oral anticoagulants, or DOACs.

CARDIO MNEMONIC FLASHCARDS

Question 118. Name five DOACs, two class suffixes, and the significance of X-a in the drug stem.

EXPANDS BREADTH OF ANTICOAGULATION OPTIONS

B_____ (Xa inhibitor)

R_____ (Xa inhibitor)

E_____ (Xa inhibitor)

A_____ (Xa inhibitor)

D_____ (Direct thrombin inhibitor)

T_____ Ten A, Xa inhibitor

H

Class suffix #1: _____

Class suffix #2: _____

395

118. DIRECT ORAL ANTICOAGULANTS (DOACs) – EXPANDS BREADTH OF ANTICOAGULATION OPTIONS

> **B**etri<u>xaban</u> (Bevyxxa)
>
> **R**ivaro<u>xaban</u> (Xarelto)
>
> **E**do<u>xaban</u> (Savaysa)
>
> **A**pi<u>xaban</u> (Eliquis)
>
> **D**abi<u>gatran</u> (Pradaxa)
>
> **T**hrombin and **T**en A, Xa inhibitors
>
> **H**

QUICK SUMMARY

DOACs are oral tablets requiring less monitoring than warfarin giving prescribers more options. Our mnemonic is EXPANDS BREADTH OF ANTICOAGULATION OPTIONS. While DOACS require less monitoring than warfarin, they have shorter half-lives, so compliance is critical. Missing a single dose can leave a patient unprotected against a clot.

B, betri<u>xaban</u> is the first oral anticoagulant effective for VTE prophylaxis and is taken once daily, versus apixaban which is taken twice daily.

R, rivaro<u>xaban</u>, is a factor Xa inhibitor with the same indications as apixaban, and dosing again varies by indication. It is dosed once daily except for over the first 21 days after VTE, during which it is dosed twice daily. Any doses over 15 mg should be taken with food. Rivaroxaban and apixaban are generally well tolerated. No antidote exists for either apixaban or rivaroxaban. As I mentioned before, there is a pipeline drug, andexanet alfa, which might be FDA approved soon that reverses Xa inhibitors and has been studied in rivaroxaban and apixaban.

Xarelto has x-a in the name for its mechanism of action blocking factor Xa [[10-"a"]], expressed as Roman numeral "X" and "a." Blocking this factor blocks coagulation. Or, you can think of banning coagulation, using the last three letters of b-a-n, ban in **rivaroxaban**. The brand name Xarelto starts with a capital "X" and little "a" again pointing to factor Xa.

E, edoxaban,

A, apixaban, is for atrial fibrillation, VTE treatment and prevention, and DVT prophylaxis after hip or knee replacement. The dosing and duration vary depending on the indication. Dosing is done twice daily and is unique for atrial fibrillation, and five milligrams twice daily is used unless the patient meets two of the following criteria: age over 80, body weight less than 60 kg, and/or serum creatinine higher than 1.5 mg/dL. Apixiban has the same "x-a-b-a-n, -xaban" stem as **rivaroxaban** and also blocks factor Xa [[10-A]].

D, dabigatran, is a direct thrombin inhibitor. It is for atrial fibrillation, VTE treatment, and DVT prophylaxis after hip replacement. We dose it twice daily. Start after 5-10 days of parenteral anticoagulation with an agent like heparin or enoxaparin. Common side effects include dyspepsia and gastritis. We take dabigatran with a full glass of water, and dispense capsules in the original container. Dabigatran has a reversal agent, **idarucizumab (Praxbind)**.

Memorize **dabigatran's** "gatran, g-a-t-r-a-n" stem to note the difference between anticoagulants. Dabi<u>gatran</u> doesn't need INR monitoring like **warfarin**, and you can note the last three letters in dabigatran as not being "I-N-R." Use the "b-i, bi" in dabigatran to remember it's a factor IIa [[two-"A"]] inhibitor.

T, thrombin and Ten A (Xa) inhibitors

H.

Next, we'll take a look at vasodilators.

Question 119. Name four vasodilators and at least three class suffixes.

HINDER ANGINA

H_____ - with suffix -d_____

I_____ M_____

N_____ - with prefix N_____

D

E

R_____ - with suffix -l_____

119. VASODILATORS – HINDER ANGINA

> **H**ydralazine (Apresoline)
> **I**sosorbide mononitrate (Imdur)
> **N**itroglycerin (NitroStat)
> **D**
> **E**
> **R**anolazine (Ranexa)

QUICK SUMMARY

The mnemonic for our vasodilators is HINDER ANGINA because these agents relieve ischemic or spastic chest pain from an oxygen demand or supply imbalance.

H, hydralazine, is a vasodilator that might cause chest pain. Hydralazine can cause a headache, hypotension, reflex tachycardia, and palpitations. See the "p-r-e-s, pres" from blood pressure in hydralazine's **Apresoline.**

I, isosorbide mononitrate, and

N, nitroglycerin, are both nitrates which work to reduce myocardial oxygen demand through nitric oxide and increase myocardial oxygen supply through increased blood flow. Isosorbide is a once or twice daily long-acting nitrate. A nitrate-free period of 10-12 hours per day prevents tolerance, so patients take doses early in the day, at least 7 hours apart. Nitroglycerin is a short-acting nitrate that is a pill, spray, or powder. Patients take at chest pain onset and can use up to three doses at 5-10 minute intervals. Nitrates can cause hypotension, dizziness, flushing, and syncope. The most common side-effect is headache. Avoid with medications for erectile dysfunction. "Nitro-" is a World Health Organization (WHO) stem. Nitroglycerin converts to nitric oxide, a vasodilator, so make sure the patient sits when he takes the med because it causes significant dizziness. Brand **Nitrostat**, takes n-i-t-r-o from "nitrous" as in a street car's extra fuel and the concern that the patient and blood pressure drop quickly or "stat, s-t-a-t" – **Nitrostat**.

D, E,

R, ranolazine, lowers intracellular calcium levels decreasing chest pain instances. With a different mechanism of action, we can use it with all the other medications to prevent chest pain. It has little to no effect on blood pressure and heart rate. It can cause a headache, dizziness, and nausea, and is typically an add-on agent for symptom control. There is a risk of QT prolongation.

Next up are other medications for prevention of stent thrombosis and reduced platelet involvement in clot formation.

Question 120. Name five drug treatment options affecting platelets, two which contain ASA and three which are ADP antagonists. Identify the P2Y12 inhibitor class stem.

ADD FOR PACT (Packed) PLATELETS

A_____ (NSAID)

D_____ / A_____ (platelet inhibitor / NSAID combination)

D

P_____ (P2Y12 inhibitor)

A_____ Antagonists (P2Y12 inhibitors)

C_____ (P2Y12 inhibitor)

T_____ (P2Y12 inhibitor)

P2Y12 inhibitor class stem: -g_____

120. Antiplatelet Agents – ADD FOR PACT (PACKED) PLATELETS

> **A**spirin (Ecotrin)
>
> **D**ipyridamole / aspirin (Aggrenox)
>
> **D**
>
> **P**rasugrel (Effient)
>
> **A**DP antagonists (P2Y12 inhibitors)
>
> **C**lopidogrel (Plavix)
>
> **T**icagrelor (Brilinta)

Quick summary

We use antiplatelet agents for ischemic heart disease, after coronary artery bypass graft (CABG), heart attack, and stroke. They can also prevent cardiac events. Our goal is to avoid platelet aggregation, so our mnemonic becomes ADD FOR PACT, P-A-C-T, PLATELETS, playing on how we pronounce packed, p-a-c-k-e-d.

A, aspirin, an irreversible prostaglandin inhibitor, is an inexpensive choice to help prevent platelet aggregation.

D, dipyridamole / aspirin, inhibits adenosine uptake into platelets, preventing aggregation. Dispense in its original container. With headache as a common side effect, we often see poor compliance. We use it after stroke. Aspirin and dipyridamole work together to prevent clots, and the brand name can be thought of as "aggregate not," **Aggrenox.**

P, prasugrel, shares the "–grel" stem with **clopidogrel.** Brand **Effient** is the word "efficient" without the "c-i," so you can think of how it may be efficient at thinning platelets.

A, ADP antagonists group (P2Y12 inhibitors), includes prasugrel, ticagrelor, and clopidogrel. The P2Y12 inhibitors bind platelet surfaces to prevent surface GPIIb/IIIa complex activation, which reduces platelet aggregation. Clopidogrel and prasugrel cause irreversible binding, while ticagrelor has reversible binding. Ticagrelor and prasugrel are both intended only for ACS patients. Clopidogrel is okay for ACS, recent stroke victims, and peripheral artery disease (PAD). Avoid prasugrel in patients with stroke history. All can increase bleeding risk, and there is no antidote. Aspirin and clopidogrel often combine as post-stroke dual antiplatelet therapy.

C, clopidogrel and aspirin work similarly, leading to a reduced likelihood that platelets will stick together and clot. **Plavix** vexes platelets, taking the "v" and "x" from vexes and p-l-a from platelets to keep the blood thin – Plavix.

T, ticagrelor, has demonstrated superiority to **clopidogrel**.

Now, we look to drugs for ST-elevated MI (STEMI).

Question 121. What four medications do you use when someone presents with a STEMI?

MOAN

M_____ is for pain

O_____ to reduce infarct size

A_____ to slow clotting and decrease clot size

N_____ for vasodilation

121. STEMI DRUGS – MOAN

> **M**orphine is for pain
>
> **O**xygen is to reduce infarct size
>
> **A**spirin (chewable) to slow clotting and decrease clot size
>
> **N**itroglycerin for vasodilation

Quick Summary

The STEMI drug mnemonic MOAN helps us hear in our minds, what a heart attack patient might cry out.

M, morphine is for pain. IV Morphine is the treatment of choice for STEMI associated pain.

O, oxygen is to reduce infarct size via nasal cannula. Recent data shows this may provide little benefit.

A, aspirin (chewable) can slow clotting and decrease clot size. It will suppress platelet aggregation, producing antithrombotic effects.

N, nitroglycerin for vasodilation can reduce cardiac pain, and reduce preload and afterload, while also increasing blood flow into the ischemic area.

Finally, we review emergency medications in general.

CARDIO MNEMONIC FLASHCARDS

Question 122. Name eight medications that one can consider an emergency drug and one emergency use for each.

LEAN AND DEMAND

L_____ for V_____ F_____

E_____ for A_____ S_____

A_____ for O_____ poisoning

N_____ for O_____ overdose

D_____ to manage C_____

E

M_____ for controlling E_____

A_____ for A_____

N

D_____ to increase B_____ P_____

122. Emergency Drugs – LEAN AND DEMAND

> **L**idocaine (Xylocaine)
> **E**pinephrine (EpiPen)
> **A**tropine (AtroPen)
> **N**aloxone (Narcan)
>
> **D**igoxin (Digitek)
> **E**
> **M**agnesium (Epsal)
> **A**miodarone (Cordarone)
> **N**
> **D**opamine (Intropin)

Quick Summary

We combine the traditional LEAN acronym with DEMAND to help us remember eight emergency medicines.

L, lidocaine, for decreasing ventricular irritability or management of ventricular fibrillation or ventricular tachycardia.

E, epinephrine, for patients showing signs and symptoms of anaphylactic shock.

A, atropine, for organophosphate poisoning or exposure to nerve agents.

N, naloxone, for patients showing apnea for unknown reasons or secondary to heroin or fentanyl overdose. It reverses suspected narcotic overdoses. It does not hurt a patient who is not experiencing an opioid OD to receive naloxone, so when health care providers approach an unconscious patient, they always give naloxone.

D, digoxin to control the ventricular rate in supraventricular tachycardia or to manage CHF. Digoxin blocks the sodium-potassium pump, and increases cardiac output and decreases heart rate. It is also for patients in later stages of heart failure when they are still symptomatic. It can control the rate of some arrhythmias. It requires therapeutic drug monitoring, as the ranges for heart failure and arrhythmias differ. It has many drug interactions with other medications. Toxicity presents as nausea, loss of appetite, blurred or yellow vision, and confusion. It can also cause low potassium and magnesium levels. Digoxin treats congestive heart failure by increasing the force of the heart's contractions. Digoxin comes from the plant *Digitalis lanata*. In Latin, *Digitalis* means something like hand or "digits," while *lanata* means "wooly" because the actual plant is fuzzy. Therefore, digoxin takes the d-i-g from *digitalis*, and the brand name **Lanoxin** takes the l-a-n from *lanata*. Alternatively, you could remember that Lanoxin and digoxin keep your heartbeat rockin'.

E,

M, magnesium, for controlling seizures associated with eclampsia during pregnancy, as a treatment for Torsade de pointes on EKG or patients with severe asthma signs and symptoms.

A, amiodarone, is for ventricular fibrillation, atrial fibrillation, ventricular tachycardia and other arrhythmias. It is a class 3 antiarrhythmic that blocks potassium channels. It is proarrhythmic. Avoid in patients without an arrhythmia. Side effects include hypotension, bradycardia, dizziness, and tremor. Rarer effects include pulmonary fibrosis, alteration in thyroid level, and vision changes. Amiodarone, like digoxin, has many drug interactions. Its half-life is exceptionally long, and it stays in the body for a while after discontinuation.

The generic amiodarone and brand **Cordarone** share the "arone" lettering. Cardiologists can also use beta-blockers, calcium channel blockers, and digoxin as antidysrhythmics.

N,

D, dopamine for increasing blood pressure associated with cardiogenic shock or the hypotensive crisis.

This look at emergency drugs wraps up our chapter and sets us up for the endocrine chapter, next.

CHAPTER 7: ENDOCRINE MNEMONIC FLASHCARDS

Question 123. Name six insulin drugs in order of duration of action and a caution with the ultra-short acting medications.

LEARN INSULIN DRUGS

L_____ (rapid-acting)

E_____ (caution with rapid-acting insulin)

A_____ (rapid-acting)

R_____ (short-acting)

N_____ (intermediate-acting)

D_____ (long-acting)

R

U

G_____ (long-acting)

123. INSULINS – LEARN INSULIN DRUGS

> **L**ispro (Humalog)
> **E**at!
> **A**spart (Novolog)
> **R**egular (Humulin R, Novolin R)
> **N**PH (Humulin N, Novolin N)
>
> **D**etemir (Levemir)
> **R**
> **U**
> **G**largine (Lantus, Toujeo, Basaglar)
> **S**

QUICK SUMMARY

We use insulin in type 1 diabetics and type 2 diabetics with poor glucose control. Insulin tells the body to store glucose in fat and muscle as glycogen. Insulins are best at lowering A1C.

Insulin side effects include weight gain and hypoglycemia. Hypoglycemia presents as dizziness, irritability, shakiness, sweating, and hunger. It's important to eat a snack containing carbohydrates when blood sugar is low.

Our mnemonic LEARN INSULIN DRUGS places insulins from shortest to longest acting with 1) lispro 2) aspart 3) regular 4) NPH 5) detemir and 6) glargine.

L, lispro, and insulin aspart are rapid-acting insulins. I use Spanish to help memorize that insulin lispro is rapid-acting. When I was in Mexico on a zip line, the person in the first tower would say, "Listo, listo" [[lee-stoh, lee-stoh]], meaning "You ready, I'm ready." Then I would fly fast down that zip cord. I replace the lispro's "s-p" with "t" because you must be "listo, listo" to take this rapid-acting insulin with a meal. Insulin lispro's brand name, **Humalog**, is a human insulin analog combining the h-u-m from human and l-o-g from analog. I picture a human on a log floating down river rapids to remember **Humalog** as rapid-acting.

E, eat, reminds us insulin lispro and aspart help control mealtime glucose spikes. Patients take those 15 minutes before meals or immediately after and they last three to five hours.

A, aspart, is the other rapid-acting insulin. Lispro and aspart are available in vials or as pens. Many use rapid-acting insulins in insulin pumps.

R, regular insulin, should begin 30 minutes before meals to control mealtime blood glucose. We can use it in IV solutions. It's technically over-the-counter, OTC, but you must go to the pharmacy to buy it. It lasts four to twelve hours and comes in vials and pens. Regular insulin is short-acting. Don't confuse this insulin with the shortest-acting insulins available, such as insulin lispro. Prescribers give **regular insulin** when patients need to adjust dosages on a sliding scale. Insulin used to come from a pig (porcine) or cow (bovine), but now matches human insulin because of molecular engineering. Therefore, the Eli Lilly brand name **Humulin** squishes the words human and insulin together, taking the h-u-m from human and u-l-i-n from insulin – Humulin.

N, NPH insulin, comes in once or twice daily doses and like regular insulin, requires no prescription. We can mix it with rapid or short-acting insulins, as long as we draw the clear insulin first. NPH lasts for 14 to 24 hours and has vial or pen forms. The "N-P-H" in **NPH insulin** stands for neutral protamine Hagedorn. The "neutral protamine" refers to how Hagedorn, the inventor, chemically altered the insulin. The "e-n" pronunciation of the letter "N" sounds a little like "i-n" and can help you remember it's an intermediate-acting insulin, taking the i-n from both intermediate and insulin – **Humulin N**.

D, detemir, is a long-acting insulin. Both **Levemir**, detemir's brand name, and Lantus, insulin glargine's brand name start with "L" for long-acting.

R, U,

G, glargine, like insulin detemir is long-acting and we give daily or sometimes twice daily. Both can last up to 24 hours. These basal insulins cover baseline glucose levels unassociated with meals throughout the day. Both detemir and glargine come as pens or vials. With **insulin glargine**, I think of the g-l-a-r in glaring and the i-n of lurking, someone who is *slowly* creeping around. For the brand **Lantus**, one of my students came up with "Lantus lasts all day long, take it at night, and your life will be prolonged." I've heard "lazy Lantus" to remember it's a 24-hour drug.

S.

Next, we take a look at biguanides and thiazolidinediones.

Question 124. Name a single biguanide, two TZDs, and their respective class suffixes.

ME DROP GLUCOSE

Me_____ - class suffix -f_____

D

R_____ - class suffix -g_____

O

P_____ - class suffix -g_____

124. Biguanides and Thiazolidinediones – ME DROP GLUCOSE

Metformin (Glucophage)

D

Rosiglitazone (Avandia)

O

Pioglitazone (Actos)

Quick Summary

A painfully grammatically incorrect mnemonic is ME DROP GLUCOSE for a biguanide and the thiazolidinediones (TZDs). Both increase insulin sensitivity to increase uptake and use of glucose in fat and muscle.

Me, metformin, a biguanide, increases insulin sensitivity like the thiazolidinediones, TZDs, and also decreases liver glucose production. We give metformin once or twice daily orally and the first agent for a newly diagnosed type-2 diabetic. It's inexpensive, lowers A1C well, and reduced micro and macrovascular complications. Metformin is weight neutral and doesn't cause hypoglycemia. Side effects may include temporary cramping, nausea, and diarrhea that do subside. Rare side effects include lactic acidosis and B_{12} deficiency. Watch for patients with renal failure and hold back metformin if patients need a contrast for a procedure.

Use the "formin, f-o-r-m-i-n" stem to remember **metformin** is a biguanide anti-diabetic. One of my students came up with this mnemonic, "If you met four men on **Glucophage**, they have diabetes then." Phagocytosis is the process of cell eating. You can use the brand name Glucophage, G-l-u-c-o-p-h-a-g-e, to think of the medication as eating, taking the g-l-u-c-o from glucose and p-h-a-g-e from phage-ing – **Glucophage**.

D,

RO, rosiglitazone, and

P, pioglitazone, are thiazolidinediones, TZDs. Chemically, they are peroxisome proliferator-activated receptor gamma (PPARgamma) agonists. They are both once-daily oral meds that don't cause hypoglycemia but do increase weight with edema. The water retention puts susceptible patients at risk for heart failure. Rare side effects include increased fracture risk and bladder cancer. These inexpensive agents have moderate A1C lowering. Pioglitazone also raises HDL and lowers triglycerides. One student said if you eat too much pie, you'll need **pie-o-glitazone** for your diabetes.

Next, we cover a sulfonylurea and meglitinide, insulin releasers.

Question 125. Name three sulfonylureas and a meglitinide and the three possible class prefixes or suffixes.

GLI GLI GLY GLUCOSE, RELEASE

Gli_____ - class prefix_____

Gli_____ - class prefix_____

Gly_____ - class prefix _____

R_____ - class suffix -g_____

ELEASE

125. Sulfonylureas and a Meglitinide
– GLI GLI GLY GLUCOSE, RELEASE

> **G**lipizide (Glucotrol)
>
> **G**limepiride (Amaryl)
>
> **G**lyburide (DiaBeta, Glynase)
>
> **R**epaglinide (Prandin)

Quick Summary

Sulfonylureas and meglitinides are insulin releasers. They stimulate pancreatic insulin secretion, decreasing blood glucose, specifically after meals. Use NSYNC's "Bye Bye Bye" song when you read GLI, GLI, GLY GLUCOSE RELEASE.

G, glipizide, comes as an extended-release formulation cleared by the liver. It works better for renally impaired patients. The "gli-, g-l-i" stem in glipizide indicates an antihyperglycemic medication. Brand **Glucotrol** implies glucose control, taking the g-l-u-c from glucose and t-r-o-l from control – **Glucotrol**.

G, glimepiride,

G, glyburide, is a metabolite the body excretes renally, so avoid in renal failure. The "gly, g-l-y" stem in **glyburide** is archaic. The stem "gli, g-l-i" replaces it in new sulfonylurea antidiabetics. The brand name **DiaBeta** combines "dia, d-i-a" from diabetic and B-e-t-a" from Beta cells, which release insulin – **DiaBeta**.

R, repaglinide, is a meglitinide that works the same as sulfonylureas, stimulating insulin secretion. Take before meals since it lowers blood glucose after eating. Like sulfonylureas, repaglinide causes hypoglycemia and weight gain. A1C reduction is modest. Brand **Prandin** includes part of "prandial" which relates to dinner or lunch, a diabetic concern.

Next, we have GLP-1 and DPP-4's.

Memorizing Pharmacology Mnemonics

Question 126. Name two GLP-1s, two class suffixes, its effect on hunger, and a dosage form. Name three DPP-4s and their class suffixes.

LESS LESS GLUCOSE

L_____ (GLP-1)

E_____ (GLP-1)

S_____ (effect on hunger)

S_____ (route of administration)

L_____ (DPP-4)

E_____ on D_____ (mechanism of action)

S_____ (DPP-4)

S_____ (DPP-4)

GLP-1 class suffix: _____

DPP-4 class suffix: _____

126. GLP-1s AND DPP-4s – LESS, LESS GLUCOSE

> Liraglutide (Victoza)
> Exenatide (Byetta, Bydureon)
> Satiety
> Subcutaneous injections
>
> Linagliptin (Tradjenta)
> Effect on DPP-4
> Sitagliptin (Januvia)
> Saxagliptin (Onglyza)

QUICK SUMMARY

GLP-1s are injectable non-insulin agents that lower glucose four ways. DPP-4 is an enzyme that breaks down GLP-1. The GLP-1's and DPP-4's have very similar mechanisms of action, so we use the same word twice in LESS, LESS GLUCOSE.

L, liraglutide, and

E, exenatide, are GLP-1 agonists that lower blood glucose by decreasing glucagon secretion, increasing insulin secretion, and slowing gastric emptying. This helps patients feel full and lose weight. GLP-1 agonists lower A1C, a patient's weight, and do not cause hypoglycemia. Side effects of GLP-1 agonists include nausea and upset stomach. As with metformin, patients tolerate this better with continued use. Liraglutide has shown cardiovascular death reduction in patients with or at high risk of CV disease. The "tide, t-i-d-e" in **exenatide** means peptide, and the "glutide, g-l-u-t-i-d-e" in **liraglutide** means a glucagon-like peptide analog or GLP.

S, satiety. Liraglutide also has an FDA weight loss approval indication.

S, subcutaneous injections. Exenatide has two formulas, one that comes twice-daily and one that's weekly. Liraglutide is only a once-daily injection. Watch for injection site reactions.

L, linagliptin, has an

E, effect on DPP-4, as does

S, sitagliptin, and

S, saxagliptin.

The "gliptin, g-l-i-p-t-i-n" stem indicates a dipeptidyl aminopeptidase-IV inhibitor, or DPP4 for short. Students associate "Lipton, L-i-p-t-o-n" iced tea sugar with **sitagliptin**. The DPP-4 inhibitors increase insulin release and lower glucagon secretion to lower blood glucose and increase satiety. We dose orally, daily, to achieve lower A1C and a little weight loss. They don't cause hypoglycemia. Compared to GLP-1 agonists, they lower A1C less and drive less weight loss. Side effects might include joint pain, new and worsening heart failure, and pancreatitis risk. Brand **Januvia** ends in "via, v-i-a" and is similar to the "dia, d-i-a" from diabetes.

Lastly, we have SGLT2 inhibitors, also used for diabetes.

Question 127. Name two SGLT2 inhibitors, their mechanism of action, and the class suffix.

EXCRETE GLUCOSE

E_____

X

C_____

RETE

Mechanism of action: _____

Class suffix: _____

127. SGLT2 Inhibitors – EXCRETE GLUCOSE

> *E*mpagliflozin (Jardiance)
> X
> *C*anagliflozin (Invokana) *(-RETE GLUCOSE)*

Quick summary

The mnemonic for SGLT2s is EXCRETE GLUCOSE because they increase urinary glucose excretion.

E, empagliflozin, and

C, canagliflozin, are SGLT2 inhibitors preventing glucose reabsorption in proximal renal tubules. This increases glucose excretion and lowers blood glucose concentrations. We give them orally once every morning. They lower A1C well, don't cause hypoglycemia, and cause weight loss. Side effects, understandably, include increased urination and thirst, and they can cause hypotension from fluid loss. With glucose in the urine, there's an increased yeast infection and UTI risk. Rare side effects include diabetic ketoacidosis and increased fracture risk. The f-l-o-z, floz (flows) in both drug names remind us to avoid in elderly patients, as increased urination and hypotension could lead to falls. Empagliflozin has been shown to reduce cardiovascular mortality.

Our next endocrine disorder involves the thyroid.

ENDOCRINE MNEMONIC FLASHCARDS

Question 128. Name a single hypothyroid medication, how you should take it, with or without food, and two hyperthyroid medications.

THYROID LAMP

L_____ (hypothyroid medication)

An E_____ S_____ (how to take)

M_____ (hyperthyroid medication)

P_____ (hyperthyroid medication)

128. Thyroid Medications – THYROID LAMP

> **L**evothyroxine (Synthroid)
> **A**n empty stomach
> **M**ethimazole (Tapazole)
> **P**ropylthiouracil (PTU)

QUICK SUMMARY

The mnemonic is THYROID LAMP, as a light switch goes up (hyperthyroid) or down (hypothyroid), so too does a patient's thyroid hormone level.

L, levothyroxine, supplements the hypothyroid patient. We diagnose hypothyroidism when low free thyroxine (T4) levels and high levels of thyroid stimulating hormone (TSH) signal an imbalance. Hypothyroid symptoms include fatigue, weight gain, cold intolerance, and weakness. Levothyroxine takes weeks to start working and requires lab monitoring. A high levothyroxine dose can create hyperthyroidism symptoms.

The generic name **levothyroxine** has the "thyro, t-h-y-r-o" from thyroid in the name. Brand **Synthroid** combines synthetic and thyroid, taking the s-y-n-t-h from synthetic and r-o-i-d from thyroid – **Synthroid**.

A, an empty stomach, with a full glass of water an hour before breakfast is ideal for levothyroxine. Often, medications and calcium supplements interact, so take this medication separately and let your doctor know of other medicines.

M, methimazole, and

P, propylthiouracil, are thionamides for hyperthyroidism, signaled by high T4 and low TSH. Symptoms include weight loss, agitation, palpitations, insomnia, and heat intolerance. Thionamides inhibit thyroid hormone synthesis. Side effects include GI upset, rash, and taste changes. We prefer propylthiouracil in a thyroid storm, but it carries a risk of liver failure. This often leads prescribers to select methimazole as their choice. The opposite is true for pregnancy. **PTU** takes "p" from propyl, "t" from "thio," and "u" from uracil in the generic **propylthio**uracil. Although "thio, t-h-i-o" suggests a sulfur atom, you can think of it as thyroid lowering, using t-h-i from thyroid and "o" from lowering – propyl*thio*uracil.

The next slide talks more about levothyroxine dosing and why monitoring levels can be essential to achieve the perfect dose.

Question 129. What is the formula for the therapeutic index?

TIE THE TD / ED

T_____

I_____

Equals

T_____

D_____

/

E_____

D_____

129. Levothyroxine Narrow Therapeutic Index – TIE the TD / ED

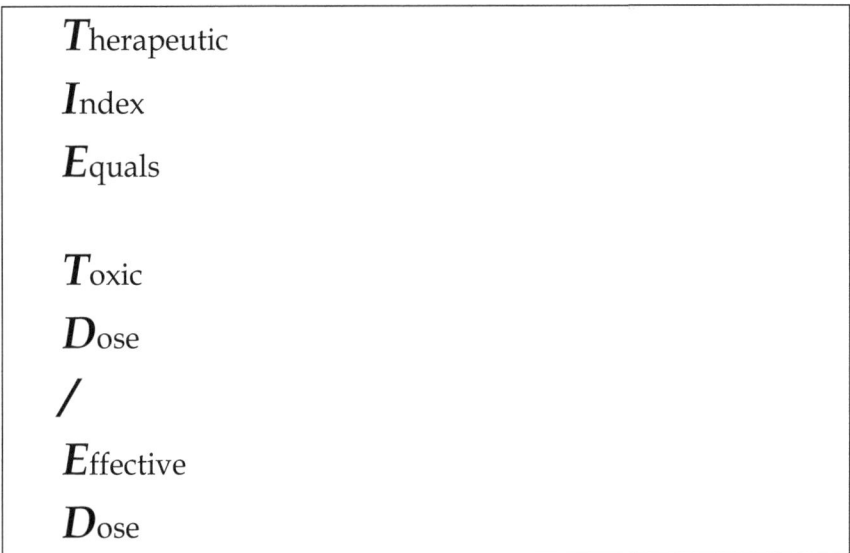

Therapeutic
Index
Equals

Toxic
Dose
/
Effective
Dose

Quick Summary

The therapeutic index, TI, involves the division of the toxic dose over the effective dose. The higher or wider the TI, the safer the drug. The lower or narrower the TI, the more unsafe the drug. If we don't properly dose levothyroxine, a narrow therapeutic index drug, patients can experience hyper or hypothyroidism.

Now, we'll move on to hormones and birth control.

Memorizing Pharmacology Mnemonics

Question 130. Write a stem for an estrogen, a progestin, and a steroid. Name four birth control products: one that adds iron, one that is triphasic, one that uses a ring, and a fourth that is a patch. Name the active ingredient in a "morning-after pill." Name a steroid that has a gel vehicle for delivery.

Hormones and birth control

1. Estrogen stem - _____

2. Progestin stem - _____

3. Steroid stem - _____

Birth control products

1. _____ - adds iron

2. _____ - triphasic

3. _____ - uses a ring

4. _____ - is a patch

Active ingredient in "morning-after pill"

_____

Steroid with gel vehicle for delivery

_____

130. HORMONES AND BIRTH CONTROL – ESTR / GEST / STER (ESTER JUST STARES)

Ethinyl Estradiol / Norethindrone /
Ferrous Fumarate (Loestrin 24 Fe)
Ethinyl Estradiol / Norgestimate (Tri-Sprintec)
Ethinyl Estradiol / Etonogestrel (NuvaRing)
Ethinyl Estradiol / Norelgestromin (OrthoEvra)

Levonorgestrel (Plan-B One-Step)

Testosterone (AndroGel)

QUICK SUMMARY

Estrogen: Ethinyl estradiol, an estrogen, inhibits FHS and ovulation, and increases aldosterone, leading to sodium and water retention. It binds free androgen, reducing acne and facial hair. Side effects include nausea and vomiting, weight gain, headache, and breast fullness. A dose that's too low can result in breakthrough bleeding. Serious adverse effects include thrombosis, which increases with higher estrogen doses. Note that when a product contains 'Lo, L-o,' it means there is a low estrogen dose, typically below 35 mcg (micrograms).

Progesterone: Norethindrone, norgestimate, etonogestrel, and norelgestromin are all progesterone types. Progesterone inhibits ovulation, thickens the uterine lining, and inhibits sperm motility. Side effects include headache and an increased appetite. Progestins vary in androgenic properties, causing

433

variable side effects. First-generation agents like norethindrone are more androgenic, causing hirsutism, hair loss, and oily skin. Norgestimate is third-generation and less androgenic.

Ethinyl estradiol / norethindrone / ferrous fumarate is monophasic. All pills contain the same hormone dose for the entire course. "Fe" is iron's chemical symbol. The estrogen stem "estr, e-s-t-r" and progestin stem "gest, g-e-s-t" help us keep the hormones straight. To remember the brand, use: "**Loestrin 24 Fe** reduces the length of menstruation, with iron supplementation, to prevent an anemic situation."

Ethinyl estradiol / norgestimate is triphasic birth control, so the hormone dose changes over a pack's course mimicking the body's natural levels. Here's another mnemonic: "**Tri-Sprintec** is triphasic; take three different doses, in seven-day spaces."

Ethinyl estradiol / etonogestrel is a vaginal ring one inserts for three weeks of each month, then removes to allow for menstruation. A student thought of **etonogestrel**'s "Etono," as in "Eat, oh no" to remember it's not an oral tablet. The brand **NuvaRing** combines "n-u" from new, "v-a" from vaginal, and "r-i-n-g" from "ring" – **Nuvaring**.

Ethinyl estradiol / norelgestromin is a transdermal patch changed weekly for three weeks, followed by a patch-free week. The patch causes higher estrogen exposure, so avoid in patients with a clot risk. Associate **norelgestromin's** "Norel" with "not oral" to remember this is a patch. Another mnemonic highlights the drug's correct placement and length of therapy: "**OrthoEvra** is a patch; put it on your arm, your abs, your buttock or back, and then take it off a week after that."

Levonorgestrel is a progestin-type agent for emergency contraception within 72 hours of unprotected intercourse. It works better the sooner it's started and often causes nausea. While it is an over-the-counter product, you find it behind the pharmacy counter because of its high price. You can recognize **levonorgestrel** as a progestin hormone product by the "gest, g-e-s-t" stem. Take it within 72 hours after sexual intercourse. It causes nausea, but flat soda can calm this down. It's now called **Plan B One-Step** because it used to take two steps or two doses to provide this contraception. I used to work in a college town pharmacy. Every Saturday and Sunday morning, I would have many students come to pick up **Plan B**. Every time, the man drove, and the woman sat in the passenger seat. When I told the male student it was fifty bucks for the Plan B, he would invariably look at her, and then pay. One time, however, I overheard her say, "Oh no, you *just* didn't." From that "just, j-u-s-t," remember the "gest, g-e-s-t" that is part of **levonorgestrel**.

Testosterone, the male hormone, treats hypogonadism. It has a parenteral injection given intramuscularly or subcutaneously, or as a topical gel, solution, and patch. The injections are painful. We apply the gel to the upper body, it is popular and patients tolerate it well. Patients should ensure that the gel dries to avoid secondary exposure to women or children while wet. Most people know the steroid hormone **testosterone**, but note that the stem for a steroid is "ster, s-t-e-r." "Andro, A-n-d-r-o" is the Greek prefix for male and gel is the vehicle in **AndroGel**, indicating it's a "male's gel."

Now, we'll look at urinary incontinence agents.

Question 131. Name a medication for urinary incontinence that is a beta-3 agonist. Identify two drugs not specific for M₃ receptors. Identify two drugs that are specific for M₃ receptors and indicate their class suffix.

LESS MODESTY NEEDED

M_____ (beta-3 agonist)

O_____ (non-specific)

D_____ - class suffix -f_____

E

S_____ - class suffix -f_____

T_____ (non-specific)

Y

131. Urinary Incontinence Agents – Less Modesty Needed

> **M**irabegron (Myrbetriq)
> **O**xybutynin (Ditropan)
> **D**arifenacin (Enablex)
> **E**
> **S**olifenacin (VESIcare)
> **T**olterodine (Detrol)
> **Y**

Quick Summary

Urinary incontinence symptoms include urgency and increase urinary frequency. Anticholinergic agents are a mainstay treatment binding muscarinic receptors which stops bladder muscle contraction. Anticholinergic side effects include dry mouth, constipation, blurry vision, and urinary retention. Elderly patients might suffer drowsiness, confusion, or blurred vision, leading to falls. Note that psychotherapy is the first-line urinary incontinence treatment as medication therapy may do more harm than good. The LESS MODESTY NEEDED mnemonic reminds us of the difficulty these patients have.

M, mirabegron, is slightly different in that it is a beta-3 agonist that relaxes detrusor muscles. Its efficacy is similar to that of anticholinergic drugs, but with less risk of dry mouth. **Myrbetriq** affects beta-3 receptors, and you can see the "be, b-e" from beta and the tri, t-r-i meaning three.

O, oxybutynin, comes as a tablet, gel, and topical patch, which is an over-the-counter patch for women, but prescription for men. One student remembered **oxybutynin** as keepin' the urin' in. Both **Di*tro*pan** and **Oxy*tro*l** have the "t-r-o" from "cont*ro*l" for cont*ro*lling an overactive bladder.

D, Darifenacin, and

S, solifenacin, are anticholinergics which are more specific for the M$_3$ bladder receptor, and cause fewer CNS effects. **Darifenacin** causes more constipation. Darifenacin and solifenacin have the same "–fenacin" stem. Both work for overactive bladder (OAB). The **Enablex** brand name hints at "enabling" the patient to "exit" the house when the OAB might have kept them in. Solifenacin's stem, "fenacin, f-e-n-a-c-i-n," should be your first clue to its function. We give solifenacin once daily, so thinking about it as "slow-fenacin" helps with remembering this point. Additionally, solifenacin solves the problem of urine that needs to be "fenced in," using the "f-e-n-c" from fenced and "i-n" to almost complete the stem "fenacin." The brand **VESIcare** contains "*vesica*, v-e-s-i-c-a," which means "bladder" in Latin, VESIcare.

T, tolterodine, is not specific for M$_3$ receptors, so it carries a CNS effects risk with its lack of selectivity. Extended release formulations like oxybutynin and tolterodine can result in less risk of dry mouth, a primary non-compliance cause. The generic name **tolterodine** has the "t-r-o" from cont*ro*l in the name, as well. The brand **Detrol** helps detrusor muscle cont*ro*l, keeping urine in, taking the d-e-t-r from detrusor and –r-o-l from control – **Detrol**.

Now, we move to agents involved in erectile dysfunction.

Question 132. Name four PDE-5 inhibitors for erectile dysfunction and the class suffix. What is a side-effect that would require an emergency room visit? What is the general mechanism of action? What drug class has a contraindication with PDE-5 inhibitors?

SAVE TONIGHT

S_____ (PDE-5)

A_____ (PDE-5)

V_____ (PDE-5)

E_____ lasting >4 hours is priapism (ER visit)

T_____ (PDE-5)

Nitric O_____, vasodilation (MoA)

N_____ are contraindicated because of a drop in
B_____ P_____

132. AGENTS FOR ERECTILE DYSFUNCTION – SAVE TONIGHT

> **S**ildenafil (Viagra)
>
> **A**vanafil (Stendra)
>
> **V**ardenafil (Levitra)
>
> **E**rection more than 4 hours is priapism and emergency
>
> **T**adalafil (Cialis)
>
> **O**xide, Nitric Oxide, NO, vasodilation MOA
>
> **N**itrates contraindicated b/c of drop in blood pressure

QUICK SUMMARY

Phosphodiesterase-5 (PDE-5) inhibitors block PDE-5 – releasing nitrous oxide, increasing cGMP, and relaxing smooth muscle – and increase genital blood flow causing an erection. Patients take sildenafil and vardenafil one hour before sexual activity, while tadalafil 30 minutes before. There is also a low-dose daily regimen. Side effects include headache, dizziness, flushing, and ocular side effects.

The erectile dysfunction mnemonic is SAVE TONIGHT because, well, you know.

The "–afil, a-f-i-l" stem specifies the P-D-E-5 inhibitor class.

S, sildenafil, is also for pulmonary arterial hypertension. There is a scene with Jack Nicholson in the movie *Something's Gotta*

Give that reminds us that patients shouldn't combine sildenafil with nitrates like **nitroglycerin** or they will end up in the emergency room. The brand **Viagra** brings viable growth, taking the v-i-a from viable and the –g-r from growth – an erection.

A, avanafil, and

V vardenafil. Vardenafil seems to have the same "-den-" substem as **sildenafil**. To levitate is to rise above the ground, so the brand name **Levitra** hints at the rising erection.

E, erection more than 4 hours is priapism and emergency, so the patient should go to the emergency department.

T, tadalafil, lasts the longest and should be taken 30 minutes before activity and in no more than one dose daily. Other indications include pulmonary arterial hypertension and BPH. I asked students how they remembered tadalafil. I'm hesitant to share their mnemonic. One said you just think of "ta-dah" as in "surprise" to remember the first two syllables "tada, t-a-d-a." I stopped them before they started on to how they remember the "fil" part of the generic name. **Cialis** is the weekend pill because it, unlike **sildenafil**, lasts through a weekend with a long half-life. Cialis is the dual bathtub commercials drug.

O, oxide, represents the nitric oxide mechanism of action.

N, nitrates contraindicated because of a drop in blood pressure. Patients should not combine nitrates and PDE-5 inhibitors, as the combination can cause severe hypotension. Interaction with nitroglycerin, nitric oxide, increases vasodilation to cause hypotensive crisis.

Now, we'll look at agents for BPH.

Question 133. Name two 5-alpha reductase inhibitors for BPH and their class suffix. Name two selective alpha-blockers for BPH and their class suffixes.

FLOWED AT LAST

F_____ class suffix -s_____

L

O

W

E

D_____ class suffix -s_____

A_____ class suffix -z_____

T_____ class suffix -o_____

133. AGENTS FOR BPH – FLOWED AT LAST

> **F**inasteride (Proscar)
> **L**
> **O**
> **W**
> **E**
> **D**utasteride (Avodart)
>
> **A**lfuzosin (Uroxatral)
> **T**amsulosin (Flomax)

QUICK SUMMARY

BPH symptoms include urinary retention and difficulty urinating. These medications allow free urine flow. Our mnemonic for benign prostatic hyperplasia, or BPH drugs, is FLOWED AT LAST.

F, finasteride is a 5-alpha-reductase inhibitor with "–steride, s-t-e-r-i-d-e" stem. The "ster, s-t-e-r" for steroid helps you remember it's for men (and the prostate). It inhibits testosterone conversion to a more active form, dihydrotestosterone (DHT) shrinking the prostate and improving symptoms. Connect brand **Proscar** to generic **finasteride**, with "That pro's car is the finest ride" – **Pros-car, finasteride**. The brand **Proscar** is for prostate care, taking the p-r-o from prostate and c-a-r from care.

Finasteride's other brand name, **Propecia**, alludes to hair growth and is meant to reverse al_opecia_ (hair loss) – pro-pecia (hair gain).

D, dutasteride and fina_steride_ both have daily dosing but can take months to be completely effective, as it shrinks the prostate slowly. Side effects include impotence, decreased libido, and ejaculation disturbances.

A, alfuzosin, and

T, tamsulosin, are selective alpha-blockers contrasting doxaz_osin_ and teraz_osin_ from the cardiology chapter which are non-selective alpha blockers. Non-selectivity causes hypotension. Selective alpha-blockers select bladder neck receptors, leading to fewer systemic adverse effects. Take once daily after the same meal as food increases absorption. Tamsul_osin_ can cause abnormal ejaculation. **Flomax** allows for a _flow_ of _max_imum urine, taking the f-l-o from flow and m-a-x from maximum – **Flomax**. The "osin" ending is not a stem, but connects **tamsul*osin*** and **alfuz*osin*** as similar.

With BPH, I think alfuzosin's brand **Uroxatral** sounds a little like "_Ur_ine con_trol_," taking the u-r from urine and "t-r-o-l, t-r-o-l" sound from "control" to match Uroxatral's t-r-a-l.

Now we'll move on to look at menopause agents.

Question 134. Name four medications prescribers might use to combat menopause symptoms.

COMPETE AGAINST MENOPAUSE

C_____ E_____

O

M_____

P_____

E_____

T

E

134. MENOPAUSE AGENTS – COMPETE AGAINST MENOPAUSE

> Conjugated estrogens (Premarin)
> O
> Medroxyprogesterone (Provera, Depo-Provera)
> Progesterone
> Estradiol (Estrace, Estraderm)
> T
> E

QUICK SUMMARY

The mnemonic is COMPETE AGAINST MENOPAUSE. These hormones help with menopause symptoms like flushing, hot flashes, mood changes, and vaginal dryness and burning.

C, conjugated estrogens, are available alone as a cream, pill, or injection, or in combination with medroxyprogesterone as an oral tablet. As with oral contraception, estr- is for estrogen replacement to help control menopausal symptoms. -gest- is for progestins. **Premarin** is for "pregnant mare's urine," the drug's source. Brands **Prempro** and **Premphase** have different estrogen / progestin levels.

O,

M, medroxyprogesterone, is available as an oral tablet or as a depot injection for contraception.

P, prog<u>es</u>terone, also provides hormonal treatment. **Progesterone** and **medroxyprogesterone** are both progestin tablets.

E, <u>estr</u>adiol, is estrogen available as a topical gel, cream, tablet, transdermal patch, and vaginal ring. Caution women starting on estrogen-containing products about their increased risk for breast cancer, clot, and endometrial cancer in those with an intact uterus who take estrogen without progestin. Recommend estrogen products at the lowest effective dose for the shortest duration, and then stop treatment. Generic **<u>estr</u>adiol** and brand names **<u>Estr</u>ace** and **<u>Estr</u>aderm** all have the e-s-t-r, <u>estr</u>ogen stem.

T,

E.

This brings us to the end of Chapter 7, Endocrine.

ALPHABETICAL LIST OF STEMS

azolam	(WHO stem) diazepam derivatives
(a)tadine	tricyclic H1 receptor antagonist
ac	anti-inflammatory agents (acetic acid derivatives)
adol	analgesics (mixed opiate receptor agonists/antagonists)
afil	phosphodiesterase type 5 (PDE5) inhibitors
alol	combined alpha and beta blockers
amivir	neuraminidase inhibitors
astine	antihistaminics (histamine-H1 receptor antagonists)
astine	H1 receptor antagonist
asvir	nonstructural protein 5A (NS5A) inhibitors
atadine	tricyclic histaminic-H1 receptor antagonists, loratadine derivatives (formerly -tadine)
azepam	antianxiety agents (diazepam type)
azolam	(WHO stem) diazepam derivatives
azoline	antihistamine/local vasoconstrictors
azosin	antihypertensives (prazosin type)
barb	barbituric acid derivatives
bendazole	anthelmintics (tibendazole type)
buvir	RNA polymerase (NS5B) inhibitors
caine	local anesthetics
cavir	carbocyclic nucleosides
cef	cephalosporins
cillin	penicillins
citabine	nucleoside antiviral / antineoplastic agents, cytarabine or azarabine derivatives
clone	hypnotics/tranquilizers (zopiclone type)
conazole	systemic antifungals (miconazole type)
coxib	cyclooxygenase-2 inhibitors
cycline	antibiotics (tetracycline derivatives)

cyclovir	antivirals (acyclovir type)
dil	vasodilators (undefined group)
dipine	phenylpyridine vasodilators (nifedipine type)
dopa	dopamine receptor agonists
dralazine	antihypertensives (hydrazine-phthalazines)
drine	sympathomimetics
dronate	calcium metabolism regulators
estr	estrogens
farin	warfarin analogs
faxine	antianxiety, antidepressant inhibitor of norepinephrine and dopamine re-uptake
fenacin	muscarinic receptor antagonists
fetamine	amfetamine derivatives
fib	fib in fibrates is not in list
fibrate	antihyperlipidemics (clofibrate type)
floxacin	fluoroquinolone (not on Stem List)
formin	hypoglycemics (phenformin type)
gab	gabamimetics
gatran	thrombin inhibitors (argatroban type)
gest	progestins
giline	Monoamine oxidase (MAO) inhibitors, type B
gli (was gly)	antihyperglycemics
gliflozin	phlorozin derivatives, phenolic glycosides
glinide	antidiabetic, sodium glucose co-transporter 2 (SGLT2) inhibitors, not phlorozin derivatives
gliptin	dipeptidyl aminopeptidase-IV inhibitors
glitazone	peroxisome proliferator activating receptor (PPAR) agonists (thiazolidene derivatives)
glutide	glucagon-like peptide (GLP) analogs
gly	antihyperglycemics
grel	platelet aggregation inhibitors, primarily platelet P2Y12 receptor antagonists
icam	anti-inflammatory agents (isoxicam type)
ifene	antiestrogens of the clomifene and tamoxifen groups

ENDOCRINE MNEMONIC FLASHCARDS

imibe	antihyperlipidaemics, acyl CoA: cholesterol acyltransferase (ACAT) inhibitors
iodarone	indicates high iodine content, antiarrhythmic
kacin	antibiotics obtained from Streptomyces kanamyceticus (related to kanamycin)
lazine	antiarrhythmic/antianginal/antihypertensive agents, phthalazine like structure
liximab	monoclonal antibodies
lizumab	monoclonal antibodies
lukast	leukotriene receptor antagonists
mab	monoclonal antibody
mantadine or mantine	antivirals/antiparkinsonians (adamantane derivatives)
mantine	antivirals/antiparkinsonians (adamantane derivatives)
melteon	selective melatonin receptor agonist
methacin	anti-inflammatory agents (indomethacin type)
micin	antibiotics (Micromonospora strains)
mycin	antibiotics (Streptomyces strain)
nal	narcotic agonists/antagonists (normorphine type)
navir	HIV protease inhibitors (saquinavir type)
nercept	tumor necrosis factor receptors
nicline	nicotinic acetylcholine receptor partial agonists/agonists
nidazole	antiprotozoal substances (metronidazole type)
nifur	5-nitrofuran derivatives
nitro	(WHO stem) NO2 derivatives
olol	beta-blockers (propranolol type)
orexant	Orexin antagonist
orphan	narcotic antagonists/agonists (morphinan derivatives)
oxacin	antibacterials (quinolone derivatives)
oxanide	antiparasitics (salicylanilide derivatives)
oxetine	antidepressants (fluoxetine type)
pamil	coronary vasodilators (verapamil type)
parin	heparin derivatives and low-molecular-weight (or depolymerized) heparins

451

peg	PEGylated compounds, covalent attachment of macrogol (polyethylene glycol) polymer
peridol	antipsychotics (haloperidol type)
peridone	antipsychotics (risperidone type)
pezil	acetylcholinesterase inhibitors used in the treatment of Alzheimer's disease
pidem	hypnotics/sedatives (zolpidem type)
pin(e)	tricyclic compounds
piprazole	(WHO stem) psychotropics, phenylpiperazine derivatives (future use is discouraged due to conflict with stem -prazole)
pitant	NK1 receptor antagonists
plon	non-benzodiazepine anxiolytics, sedatives, hypnotics
prazole	antiulcer agents (benzimidazole derivatives)
pred	prednisone and prednisolone derivatives
previr	serine protease inhibitors
pril	antihypertensives (ACE inhibitors)
prim	antibacterials (trimethoprim type)
profen	anti-inflammatory/analgesic agents (ibuprofen type)
prost	prostaglandins
racetam	nootropic agents (learning, cognitive enhancers), piracetam type
rifa	antibiotics (rifamycin derivatives)
sal	anti-inflammatory agents (salicylic acid derivatives)
sartan	angiotensin II receptor antagonists
semide	diuretics (furosemide type)
setron	serotonin 5-HT3 receptor antagonists
sidone	antipsychotic with binding activity on serotonin (5-HT2A) and dopamine (D2) receptors
spirone	anxiolytics (buspirone type)
ster	steroids (androgens, anabolics)
steride	testosterone reductase inhibitors
stigmine	cholinesterase inhibitors (physostigmine type)
sulfa	antimicrobials (sulfonamides derivatives)

ENDOCRINE MNEMONIC FLASHCARDS

tacept	T-cell receptors
tegravir	integrase inhibitors
terol	bronchodilators (phenethylamine derivatives)
thiazide	diuretics (thiazide derivatives)
thromycin	macrolide (not on Stem List)
tiapine	antipsychotics (dibenzothiazepine derivatives)
tiazem	calcium channel blockers (diltiazem type)
tide	peptides
tidine	H2-receptor antagonists (cimetidine type)
toin	antiepileptics (hydantoin derivatives)
traline	selective serotonin reuptake inhibitors (SSRI)
trexate	antimetabolites (folic acid derivatives)
triptan	antimigraine agents (5-HT1 receptor agonists); sumatriptan derivatives
triptyline	antidepressants (dibenzol[a.d.]cycloheptane derivatives)
trop(ine)	atropine derivatives; Subgroups: tertiary amines (e.g., benztropine)
trop(ium)	atropine derivative (quaternary ammonium salt)
uracil	uracil derivatives used as thyroid antagonists and as antineoplastics
vastatin	antihyperlipidemics (HMG-CoA inhibitors)
vir	antiviral
virenz	non-nucleoside reverse transcriptase inhibitors; benzoxazinone derivatives
viroc	CC chemokine receptor type 5 (CCR5) antagonists
virtide	antiviral peptides
vudine	antineoplastics; antivirals (zidovudine group) (exception: edoxudine)
xaban	antithrombotics, blood coagulation factor XA inhibitors
xostat	xanthine oxidase/dehydrogenase inhibitors
zolamide	carbonic anhydrase inhibitors
zolid	oxazolidinone antibacterials

Generic and Brand Index

abaloparatide, 110
abatacept, 105
Abilify, 310
Abreva, 227
ABSSSI, 193
Accolate, 159
Accupril, 340
acetaminophen, 6, 81, 84, 91, 220, 243, 257
Acetaminophen, 81
acetylcysteine, 257
Achromycin, 202, 214
AcipHex, 24
aclidinium, 155
Actonel, 108
Actos, 417
acyclovir, 227
adalimumab, 105
Adderall, 265
Advair, 152
Advil, 73
Afrin, 129
Aggrenox, 403
akathisia, 308
albuterol, 147, 150, 155
Aldactone, 22, 367, 371
aldosterone, 338
alendronate, 108
Aleve, 73, 75, 77
alfuzosin, 444
alirocumab, 383
aliskiren, 338
Alka-Seltzer, 13, 16

Allegra, 125
allopurinol, 116
almotriptan, 99
Aloxi, 57
alprazolam, 249, 253
Altace, 340
aluminum hydroxide, 13
amantadine, 323
Amaryl Amaryl, 420
Ambien, 243
Amerge, 99
amikacin, 196
Amikin, 196
amiodarone, 394, 408
Amitiza, 66
amitriptyline, 285
amlodipine, 364
amoxicillin, 6, 32, 35, 163, 164, 167
amphetamine, 265
Amphojel, 13
amphotericin, 222
ampicillin, 163
Anbesol, 330
Ancobon, 222
Androgel, 433
angioedema, 340
angiotensin I, 337
angiotensin II, 338
angiotensinogen, 337
Antabuse, 38
anticholinergic, 49, 121
Antivert, 121

455

APAP, 81
apixaban, 396
aprepitant, 57
Apresoline, 400
ARBs, 346
Aricept, 327
aripiprazole, 310
Arixtra, 389
Asmanex, 144
aspart, 413
aspirin, 73, 75, 375, 394, 403, 406
aspirin / acetaminophen / caffeine, 81
aspirin / dipyridamole, 403
Astelin, 123
Atacand, 346
atenolol, 352, 357
Ativan, 249, 253, 255
atomoxetine, 265
atorvastatin, 377
ATRIPLA, 233
atropine, 46, 408
Atrovent, 155
Augmentin, 167
autoinducer, 296
avanafil, 440
Avandia, 417
Avapro, 346
Avelox, 192
Avodart, 444
Axert, 99
Axid, 9, 18
Azactam, 176
azathioprine, 102
azelastine, 123
Azelastine, 123
azithromycin, 202, 210
aztreonam, 176
Azulfidine, 66, 102
Bacitracin, 196

baclofen, 112
Bactocil, 167
Bactrim, 202, 394
Basaglar, 413
Baxdela, 192
beclomethasone, 144
Belsomra, 243
Benadryl, 121
benazepril, 340
Benemid, 116
Benicar, 346
Bentyl, 66
benzocaine, 330
benzonatate, 136
benztropine, 323
beta-lactam ring, 167
betrixaban, 396
Bevyxxa, 396
Biaxin, 32, 35, 210
Bicillin, 163
bismuth subsalicylate, 35, 46
Boniva, 108
BPH, 349
bradykinin, 340
Breo Ellipta, 152
Brevibloc, 352
brexpiprazole, 310
Brillinta, 403
Bromfed, 121
brompheniramine, 121
budesonide, 133, 144, 152
buprenorphine / naloxone, 94
bupropion, 262
Buspar, 249
buspirone, 249
butalbital / acetaminophen / caffeine, 81
Bydureon, 423
Byetta, 423
Bystolic, 352
Calan, 361

calcium carbonate, 13
canagliflozin, 426
candesartan, 346
Capoten, 340
captopril, 340
Carafate, 41
carbamazepine, 296, 314
carbapenems, 178
Cardene, 364
Cardizem, 361
Cardura, 349
Cardura XL, 150
carisoprodol, 112
carteolol, 357
Cartrol, 357
carvedilol, 352, 357
Catapres, 265
Cayston, 176
CCBs, 361
cefazolin, 171
cefdinir, 171
cefepime, 171
Cefobid, 38
cefoperazone, 38
Cefotan, 38
cefotetan, 38
ceftaroline, 171
ceftazadime, 171
Ceftin, 171
ceftriaxone, 171
cefuroxime, 171
Celebrex, 75
celecoxib, 75
Celexa, 272, 274
cephalexin, 171
cetirizine, 125
Chantix, 262
Cheratussin AC, 136
Chlorphen, 121
chlorpheniramine, 121

chlorpheniramine polistirex, 136
chlorpromazine, 304
cholinergic, 51
Cialis, 440
cimetidine, 9, 18, 22, 394
Cimetidine, 22
CINV, 56, 58
Cipro, 192, 394
ciprofloxacin, 192, 394
citalopram, 272
Clarinex, 125
Clarispray, 133
clarithromycin, 32, 35, 210
Claritin, 125
Claritin-D, 129
clavulanate, 167
Cleocin, 202
clindamycin, 202
Clinoril, 73, 75
clonazepam, 249, 253
clonidine, 265
clopidogrel, 403
Clostridium difficile, 30, 181
clotrimazole, 222
clozapine, 310
Clozaril, 310
cocaine, 330
codeine, 91, 136
Cogentin, 323
Colace, 53, 54
colchicine, 116
Colcrys, 116
colesevelam, 385
Combivent, 155
Compazine, 63
Comtan, 323
concentration-dependent, 188
Concerta, 265
conjugated estrogens, 447
Cordarone, 394, 408

Coreg, 352, 357
Corgard, 357
Coumadin, 389
Cozaar, 346
Crestor, 377
Cubicin, 188
cyclobenzaprine, 112
Cymbalta, 282
Cytotec, 41
dabigatran, 396
daclatasvir, 239
Daklinza, 239
dalbavancin, 185
Dalvance, 185
dapsone, 394
Daptomycin, 188
darifenacin, 437
darunavir, 233
Daytrana, 265
DEA schedule, 84
Decadron, 57, 139
delafloxacin, 192
Deltasone, 139
Demerol, 84
Depakote, 296, 314
DepoProvera, 447
depression diagnosis, 269
desloratadine, 125
desvenlafaxine, 282
detemir, 413
Detrol, 437
dexamethasone, 57, 139
Dexilant, 24
dexlansoprazole, 24
dexmethylphenidate, 265
dextroamphetamine, 265
diazepam, 112, 253
diclofenac, 73
dicloxacillin, 167
dicyclomine, 66
Dificid, 210

Diflucan, 222, 394
Digibind, 257
digoxin, 257, 408
Dilantin, 314
diltiazem, 361
Dimetapp, 121
Diovan, 346
diphenhydramine, 121, 243
diphenoxylate, 46
Disalcid, 73
disulfiram, 38
Disulfiram, 38
Ditropan, 437
divalproex, 314
DOACs, 396
dobutamine, 150
Dobutrex, 150
docosanol, 227
docusate sodium, 53
Dolophine, 84
donepezil, 327
dopamine, 408
Doribax, 178
doripenem, 178
Doryx, 214
dorzolamide, 371
doxazosin, 150, 349
doxepin, 285
doxycycline, 35, 214
DPP-4, 423
Dulera, 152
duloxetine, 282
Duoneb, 155
Duragesic, 84
dutasteride, 444
Dyazide, 367
Dynacirc, 364
Dynapen, 167
Dyrenium, 367
dystonia, 308
echinocandins, 222

Index

Ecotrin, 73, 75, 76, 394, 403
edoxaban, 396
efavirenz, 233
Effexor, 282
Effient, 403
Elavil, 285, 286
elbasvir, 239
Eldepryl, 291, 294, 323
eletriptan, 99
Eliquis, 396
Emend, 57
Emend injectable, 57
empagliflozin, 426
Emsam, 291, 294
Emtricitabine, 233
Emtriva, 233
E-mycin, 394
E-Mycin, 41
Enablex, 437
enalapril, 340, 375
enalaprilat, 341
Enbrel, 105
enfuvirtide, 233
enoxaparin, 389
entacapone, 323
Enulose, 55
epinephrine, 159, 408
EpiPen, 159, 160
EPS, 304
Epsal, 408
ertapenem, 178
EryPed, 210
erythromycin, 41, 210, 394
escitalopram, 272
esmolol, 352
esomeprazole, 24, 32, 35
estazolam, 243
Estrace, 447
Estraderm, 447
estradiol, 433, 447
estrogens, 22

eszopiclone, 243, 245
etanercept, 105
ethambutol, 217
etodolac, 75
etonogestrel, 433
evolucumab, 383
Excedrin Migraine, 81
Exelon, 327
exenatide, 423
ezetemibe, 385
FAB-4, 394
famotidine, 9, 18
febuxostat, 116
felodipine, 364
fenofibrate, 385
fenoldopam, 250
fentanyl, 84
ferrous fumarate, 433
fexofenadine, 125
fidaxomicin, 210
finasteride, 444
Fioricet, 81
Flagyl, 35, 38, 189, 394
Flexeril, 112
Flomax, 444
Flonase, 133
Flovent, 144
Flovent HFA, 145
fluconazole, 222, 394
flucytosine, 222
flumazenil, 257
fluoxetine, 272
fluticasone, 144
fluticasone furoate, 133, 152
fluticasone propionate, 133, 152
fluvastatin, 377
fluvoxamine, 272
Focalin, 265
fondaparinux, 389
formoterol fumarate, 152

459

formoterol furoate, 152
Fortaz, 171
Forteo, 110
Fosamax, 108
fosaprepitant, 57
fosinopril, 340
Frova, 99
frovatriptan, 99
Fungizone, 222
furosemide, 367
Fuzeon, 233, 234
gabapentin, 319
Gabitril, 319
galantamine, 327
Garamycin, 196, 197
gemfibrozil, 385
gentamicin, 196
Geodon, 310
glargine, 413
glimepiride, 420
glipizide, 420
GLP-1s, 423
Glucophage, 417
Glucotrol, 420
glyburide, 420
glycerin suppository, 55
Glynase, 420
Gocovri, 323
golimumab, 105
Grifulvin, 222
Grifulvin V, 38
griseofulvin, 38, 222
Gris-PEG, 222
guaifenesin, 136
guaifenesin /
   dextromethorphan, 136
Halcion, 243, 253
Haldol, 304
haloperidol, 304
Helicobacter pylori, 30
heparin, 257, 389

hepatotoxic medications, 220
Humalog, 413
Humira, 105
Humulin N, 413
hydralazine, 400
hydrochlorothiazide, 367
hydrocodone, 84, 136
*H*ydrocodone / Ibuprofen, 84
hydrocortisone, 69
Hydrodiuril, 367, 371
hydroxychloroquine, 102
hydroxyzine, 121
hyperkalemia, 340
Hytrin, 349
ibandronate, 108
ibuprofen, 73, 75, 375
idarucizumab, 257, 398
Imdur, 400
imipenem, 178
imipramine, 285
Imitrex, 99
Imodium, 46, 47
Imuran, 102
Incruse Ellipta, 155
Inderal, 150, 357
Inderal LA, 352
Indocin, 75, 116, 394
indomethacin, 75, 116, 394
infliximab, 66, 105
INH, 217, 220
Intropin, 408
Invanz, 178
Invega, 310
Invokana, 426
ipratropium, 155
irbesartan, 346
Isentress, 233
isocarboxazid, 291, 294
isoniazid, 217, 220
isosorbide mononitrate, 400
isradipine, 364

itraconazole, 222
Januvia, 423
Jardiance, 426
Kadian, 84
Kapvay, 265
Keflex, 171
Kefzol, 171
Keppra, 319
ketoprofen, 75
ketorolac, 75
Klonopin, 249, 253
Krystexxa, 116
labetalol, 352
lactulose, 55
Lamictal, 296, 319, 320
Lamisil, 222
lamotrigine, 296, 319
lansoprazole, 24
Lantus, 413
Lasix, 367, 371
Lescol, 377
levalbuterol, 147
Levaquin, 192
Levemir, 413
levetiracetam, 319
Levitra, 440
levocetirizine, 125
levodopa / carbidopa, 323
levofloxacin, 192
levonorgestrel, 433
levothyroxine, 428
Lexapro, 272
lidocaine, 330, 408
linaclotide, 68
linagliptin, 423
linezolid, 208
Linzess, 68
Lioresal, 112
Lipitor, 377
liraglutide, 423
lisdexamfetamine, 265

lisinopril, 340, 375
lispro, 413
lithium, 296, 301
Lithobid, 296
Lodine, 75
Loestrin 24 FE, 433
Lomotil, 46
loperamide, 46
Lopid, 385
Lopressor, 352, 357
loratadine, 125, 129
lorazepam, 249, 253, 255
losartan, 346, 375
Lotensin, 340
lovastatin, 377
Lovaza, 385
Lovenox, 389
lubiprostone, 66
luliconazole, 222
Luminal, 259
Lunesta, 243
Luvox, 272
Luzu, 222
Lyrica, 319, 320
Macrobid, 202
Macrodantin, 202
magnesium, 408
magnesium hydroxide, 13
mannitol, 367
MAOIs, 291
maraviroc, 233
Marplan, 291, 294
Matulane, 38
Maxalt, 99
Maxipime, 171
Maxzide, 367
meclizine, 121
Medrol, 139
medroxyprogesterone, 447
meloxicam, 75
memantine, 327

meperidine, 84
Mephyton, 257
mero_penem_, 178
Merrem, 178
me_sa_lamine, 66
Metadate, 265
Metamucil, 53
metaxalone, 112
met_formin_, 417
methadone, 84
meth_icillin_, 167
methimazole, 428
methocarbamol, 112
metho_trexate_, 102
methyl_dopa_, 280
methyl_naltrexone_, 97
methylphenidate, 265
methyl_prednisolone_, 139
metoclopramide, 41
meto_prolol_, 150, 352, 357
meto_prolol_ succinate, 354
metoprolol tartrate, 354
metro_nidazole_, 35, 38, 394
Mevacor, 377
mid_azolam_, 249, 253
Milk of Magnesia, 13
Minipress, 349
Minocin, 214
mino_cycline_, 214
miosis, 89
mirabegron, 437
MiraLAX, 53
Mirapex, 323
mirtazapine, 285
miso_prost_ol, 41
Mobic, 75
mometasone, 133, 144, 152
Monopril, 340
monte_lukast_, 159
morphine, 84, 406
Motrin, 73, 75

Movantik, 97
Moxatag, 32, 35, 163
moxi_floxacin_, 192
MRSA, 169
MS Contin, 84
MSSA, 169
mu receptor, 89
Mucinex DM, 136
Mucomyst, 257
Myambutol, 217
Mycelex, 222
Mycobacterium tuberculosis, 217
Mycobutin, 217
Mycostatin, 222
Myrbetriq, 437
nabumetone, 75
nad_olol_, 357
naf_cillin_, 167
na_ldemedine_, 97
na_loxegol_, 97
na_loxone_, 94, 257
Namenda, 327
naproxen, 73, 75, 375
nara_triptan_, 99
narcan, 408
Narcan, 94, 257
Nardil, 291, 294
Nasonex, 133
nebivolol, 352
neo_mycin_ / polymyxin B, 196
Neosporin, 197
NeoSynephrine, 129
Neurontin, 319
Nexium, 24, 26, 32, 35
niacin, 385
Niaspan, 385
nicar_dipine_, 364
nife_dipine_, 364
nitrofuran_toin_, 202
nitroglycerin, 400, 406

NitroStat, 400
nizatidine, 9, 18
norelgestromin, 433
norethindrone, 433
norgestimate, 433
Normodyne, 352
nortriptyline, 285
Norvasc, 364
Novolog, 413
Noxafil, 222
NPH, 413
NSAIDs, 73
NuvaRing, 433
nystagmus, 317
nystatin, 222
olanzapine, 310
olmesartan, 346
Olysio, 239
omalizumab, 159
ombitasvir, 239
omega-3-acid ethyl esters, 385
omeprazole, 24
Omnicef, 171
ondansetron, 57
Onglyza, 423
opioid antagonists, 94
OraPred, 139
OraVerse, 150
Orbactiv, 185
Orencia, 105
oritavancin, 185
orphenadrine, 123
OrthoEvra, 433
Orudis, 75
oseltamivir, 230
Osmitrol, 367, 371
oxacillin, 167
oxazepam, 255
oxcarbazepine, 319
oxybutynin, 437
oxycodone, 84

Oxycontin, 84
oxygen, 406
OxyIR, 84
oxymetazoline, 129
P2Y12 inhibitors, 403
paliperidone, 310
palivizumab, 237
palonosetron, 57
Pamelor, 285
pantoprazole, 24
Parkinsonism, 308
Parnate, 291, 294
paroxetine, 272
Paxil, 272
PCSK9 inhibitors, 383
PDE-5, 440
pegloticase, 116
penicillin, 6, 8
penicillin G, 163
penicillin VK, 163
Pentasa, 66
Pentothal, 259
Pepcid, 9, 18
Pepto-Bismol, 35, 46
peramivir, 230
Percocet, 84
phenelzine, 291, 294
Phenergan, 63
phenobarbital, 259
phentolamine, 150
phenylephrine, 129
phenytoin, 314
physostigmine, 257
phytonadione, 257
pioglitazone, 417
Plan-B One- Step, 433
Plaquenil, 102
Plavix, 403
plecanatide, 68
polyethylene glycol, 53
posaconazole, 222

463

Pradaxa, 257, 396
Praluent, 383
pramipexole, 323
Prandin, 420
prasugrel, 403
Pravachol, 377
pravastatin, 377
Praxbind, 257, 398
prazosin, 349
prednisolone, 139
prednisone, 139
pregabalin, 319
Premarin, 447
Prevacid, 24
Prezista, 233
Prilosec, 24, 26
Primaxin IV, 178
Principen, 163
Pristiq, 282
Proair, 147
ProAir, 150
ProAir HFA, 148
probenecid, 116
procarbazine, 38
Procardia, 364, 365
prochlorperazine, 63
progesterone, 447
promethazine, 63
propranolol, 150, 280, 352, 357
Propranolol, 353
propylthiouracil, 428
Proscar, 444
ProSom, 243
protamine sulfate, 257
Protonix, 24
protriptyline, 285
Proventil, 147
Provera, 447
Prozac, 272
pseudoephedrine, 129
psyllium, 53

PTU, 428
Pulmicort, 144
pyrazinamide, 217, 220
PZA, 217, 220
quetiapine, 310
Quillivant, 265
quinapril, 340
Qvar, 144
rabeprazole, 24
raltegravir, 233
ramelteon, 243
ramipril, 340
Ranexa, 400
ranitidine, 9, 18
ranolazine, 400
Rapivab, 230
Razadyne, 327
Reclast, 108
red man syndrome, 181
Reglan, 41
regular, 413
Relafen, 75
Relenza, 230
Relistor, 97
Relpax, 99, 100
Remeron, 285
Remicade, 66, 105, 237
renin, 337
repaglinide, 420
Repatha, 383
Requip, 323
Restoril, 243, 253, 255
Rexulti, 310
rhabdomyolysis, 381
Rheumatrex, 102
Rhinocort, 133
rifabutin, 217
Rifadin, 217
rifampin, 217
risedronate, 108
Risperdal, 22, 310

Index

risperidone, 22, 310
Ritalin, 265
rivaroxaban, 396
rivastigmine, 327
rizatriptan, 99
Robaxin, 112
Robitussin DM, 136
Rocephin, 171
Romazicon, 257
ropinirole, 323
rosiglitazone, 417
rosuvastatin, 377
Rozerem, 243
RSV, 237
safinamide mesylate, 323
salmeterol, 147, 152
salsalate, 73
Savaysa, 396
saxagliptin, 423
scopolamine, 57
Secnidazole, 190
secobarbital, 259
Seconal, 259
selegiline, 291, 294, 323
Selzentry, 233, 234
senna, 53
Senokot, 53
Serax, 255
Seroquel, 310
sertraline, 272
Severent, 147
SGLT2s, 426
sildenafil, 440
Silenor, 285
simeprevir, 239
Simponi, 105
simvastatin, 377, 394
Sinemet, 323
Singulair, 159
sitagliptin, 423
Sivextro, 202, 208

Skelaxin, 112
SNRIs, 282
sodium bicarbonate, 13
sofosbuvir, 239
Solarcaine, 330
solifenacin, 437, 438
Solosec, 190
Soma, 112, 243
Sonata, 243
Sovaldi, 239
Spiriva, 155
spironolactone, 22, 367
Sporonox, 222
SSRI side effects, 277
SSRIs, 272
Stendra, 440
Stevens-Johnson Syndrome, 205
Strattera, 265, 267
streptomycin, 196
Sublimaze, 84
Suboxone, 94
sucralfate, 41
Sudafed, 129
sulfamethoxazole, 202, 394
Sulfamethoxazole, 205
sulfasalazine, 66, 102
sulindac, 73, 75
sumatriptan, 99
Sustiva, 233
suvorexant, 243
Symbicort, 152
Symproic, 97
Synagis, 237
syncope, 349
Synthroid, 428
tadalafil, 440
Tagamet, 9, 18, 22, 394
Tamiflu, 230
tamsulosin, 444
Tapazole, 428

465

tardive dyskinesia, 304, 308
Tazicef, 171
TCA, 285
tedizolid, 202, 208
Teflaro, 171
Tegretol, 296, 314
Tekturna, 338
telavancin, 185
temazepam, 243, 253, 255
tenofovir, 233
Tenormin, 352, 357
Terazol 7, 222
terazosin, 349
terbinafine, 222
terconazole, 222
terfenadine, 127
Teriparatide, 110
Tessalon Perles, 136
testosterone, 433
tetracycline, 202, 214
therapeutic index, 431
thiopental, 259
Thorazine, 304
tiagabine, 319
ticagrelor, 403
time-dependent, 188
tiotropium, 155
tizanidine, 112
tobramycin, 196
Tobrex, 196
Tofranil, 285
tolterodine, 437
Topamax, 319
topiramate, 319
Toprol XL, 150, 352, 357
Toradol, 75
Toujeo, 413
Tradjenta, 423
tramadol, 91
Transderm-Scop, 57
tranylcypromine, 291, 294

trazodone, 243
triamterene, 367
triazolam, 243, 253
Tricor, 385
Trileptal, 319
trimethoprim, 202, 394
Trimethoprim, 205
Tri-Sprintec, 433
Trulance, 68
Trusopt, 371
Tudorza, 155
Tums, 13
Tussionex, 136
Tylenol, 81, 91, 220, 243
Tylenol #2, 92
Tylenol #3, 91
Tylenol #4, 92
Tylenol PM, 243
Tymlos, 110
tyramine, 292
Uloric, 116
Ultracet, 91
Ultram, 91
umeclidinium, 155
Unipen, 167
Uroxatral, 444
valacyclovir, 227
Valium, 112, 253
valproic acid, 296
valsartan, 346, 375
Valtrex, 227
Vancocin, 181
vancomycin, 181
Vancomycin, 185
vardenafil, 440
varenicline, 262
varicella, 227
Vasotec, 340
Veetids, 163
venlafaxine, 282
Ventolin, 147

Veramyst, 133
verapamil, 361
Versed, 249, 253
Vesicare, 437
Vfend, 222
Viagra, 440
Vibativ, 185
Vicodin, 84
Vicoprofen, 84
Victoza, 423
Viibryd, 275
vilanterol, 147, 152
vilazodone, 275
Viread, 233
Vistaril, 121
Voltaren, 73, 75
voriconazole, 222
VRE, 208
Vyvanse, 265
warfarin, 389
Welchol, 385
Xa inhibitors, 396
Xadago, 323
Xanax, 249, 253
Xarelto, 396
Xolair, 159, 160, 237
Xopenex, 147
Xylocaine, 330

Xyzal, 125
zafirlukast, 159
Zaleplon, 243
Zanaflex, 112
zanamivir, 230
Zantac, 9, 18
Zepatier, 239
Zestril, 340
Zetia, 385
ziprasidone, 310
Zithromax, 202, 210
Zocor, 377, 394
Zofran, 57
zoledronic acid, 108
zolmitriptan, 99
Zoloft, 272
zolpidem, 243
Zometa, 108
Zomig, 99
Zonegram, 319
zonisamide, 319
zoster, 227
Zovirax, 227
Zyban, 262
Zyloprim, 116
Zyprexa, 310
Zyrtec, 125
Zyvox, 208

Index

Made in the USA
Thornton, CO
05/04/22 19:08:24

5be142b5-bb00-460c-847f-2dc3d6640fceR02